NATIONAL PARKS FOREVER

NATIONAL PARKS

FOREVER

Fifty Years of Fighting and a Case for Independence

JONATHAN B. JARVIS
& T. DESTRY JARVIS

The University of Chicago Press   CHICAGO AND LONDON

The University of Chicago Press, Chicago 60637
The University of Chicago Press, Ltd., London
© 2022 by Jonathan B. Jarvis and T. Destry Jarvis
Published 2022
Printed in the United States of America

31 30 29 28 27 26 25 24 23 22   1 2 3 4 5

ISBN-13: 978-0-226-81909-9 (cloth)
ISBN-13: 978-0-226-81908-2 (paper)
ISBN-13: 978-0-226-81910-5 (e-book)
DOI: https://doi.org/10.7208/chicago/9780226819105.001.0001

Library of Congress Cataloging-in-Publication Data

Names: Jarvis, Jonathan B., author. | Jarvis, T. Destry, author.
Title: National parks forever : fifty years of fighting and a case for independence /
    Jonathan B. Jarvis and T. Destry Jarvis.
Description: Chicago : University of Chicago Press, 2022. | Includes bibliographical
    references and index.
Identifiers: LCCN 2021041762 | ISBN 9780226819099 (cloth) | ISBN 9780226819082
    (paperback) | ISBN 9780226819105 (ebook)
Subjects: LCSH: United States. National Park Service—History. | National parks and
    reserves—United States. | Conservation of natural resources—Government policy—
    United States.
Classification: LCC SB482.A4 J37 2022 | DDC 363.6/80973—dc23
LC record available at https://lccn.loc.gov/2021041762

♾ This paper meets the requirements of ANSI/NISO Z39.48-1992
(Permanence of Paper).

This book is dedicated to the men and women of the National Park Service who devote their careers to the stewardship of our national parks so that they may be enjoyed, unimpaired, by future generations.

And to our wives, Paula and Barbara, who have tolerated our incessant discussions about national parks for more than forty years.

CONTENTS

FOREWORD

Millions of us have a national park story. My story began as a rambunctious five-year-old surfing in the backseat of my grandfather's 1956 Buick Roadmaster while he careened up a winding road along the Rogue River. The car felt like a rolling ship in heavy seas as we headed to our favorite national park, Crater Lake.

My sea legs were grounded when I jumped out of the Roadmaster and took in Wizard Island, surrounded by the bluest blue water I had ever seen. Chipmunks and golden-mantled ground squirrels provided my first national park wildlife experience. Eager for a snack, they gathered around us as we were having a picnic—perched on the crater rim with a clear view of the deep, liquid-filled caldera that was surely a natural wonder of the world. On that warm summer day in 1956 I fell in love with what I was sure was the most beautiful place on earth.

Crater Lake continued to draw me back. When I came home to introduce Elizabeth, my wife-to-be, to my parents, I could not wait to jump in the car and drive her the ninety miles up that twisty road to share my first national park experience. We strapped on our snowshoes and explored Crater Lake under a clear cobalt-blue December sky. I regaled Elizabeth with the story of riding my bicycle around the thirty-three-mile crater rim road. We had a picnic where my family had picnicked thirty-three years before, and we talked about wanting to share places like this with the children we hoped to have.

True to our wish, with our three children, we have explored Crater Lake, Yellowstone, Yosemite, Sequoia, Shenandoah, Grand Teton, Glacier, Rocky

Mountain, Redwoods, Rainier, Point Reyes, Hawai'i Volcanoes, Everglades, Olympic, and Theodore Roosevelt National Parks. The list continues on.

We have slept under the stars in African parks where this grand national park experiment was exported to create Kruger National Park, Serengeti National Park, Kalahari Gemsbok National Park, Virunga National Park, Etosha National Park, Amboseli National Park, and many more.

During our journeys, we have met inspiring people dedicated to taking care of the parks we cherish—with an unwavering attentiveness to both immediate needs and a commitment to preservation for future generations. For example, there was a Yakutat District Ranger, Clarence Summers, who rappelled off cliffs with me above the breathtaking Wrangell-St. Elias landscape. Clarence went beyond the call of duty in helping me secure camera traps to photograph surges of the park's seventy-five-mile-long Hubbard Glacier. In Yellowstone, senior wildlife biologist Doug Smith gave us an appreciation of the critical role wolves play in the ecosystem. Both mitigating human conflict with wildlife and the necessity of listening to and collaborating with local communities were constant topics of conversation. We were always impressed with the breadth of knowledge and passion park staff brought to their jobs every day. The same held true for the 400,000 National Park Service volunteers. Elizabeth, our children, and I have developed lifelong friendships with those who share our deep connection to and love of these special places. We have also become acutely aware that national parks demand our attention and should never be taken for granted.

That is what this book is about—paying attention and never taking national parks for granted. No two people are better qualified to celebrate our national parks than Jonathan Jarvis and Destry Jarvis. They can address the challenges facing the parks and advocate for thoughtful, bold steps forward. Between the two of them they have more than ninety years of experience with our parks. No one knows the landscape, inside and out, like the Jarvis brothers.

For more than two years, when I was editor-in-chief of *National Geographic* magazine, I witnessed the expertise and dedication of Jonathan and Destry as we collaborated on our Yellowstone National Park project. With Jonathan's support (he was then the eighteenth director of the National Park Service), Destry (former assistant director of the NPS) and I met with Dan Wenk, the superintendent of Yellowstone, to begin planning articles that would celebrate the hundredth anniversary of the National Park Service by exploring the wonders and the worries within and surrounding the world's first national park. Our working relationship was grounded by *National Geographic*'s steadfast commitment to the creation of the National Park Service back in 1916.

One hundred years later, the May 2016 magazine was solely focused on Yellowstone. It became one of the most popular issues we ever published. No surprise! More than four million people visited Yellowstone in 2019. Total national park visits now exceed more than 320 million people per year . . . visits to not just the grand western parks such as Yellowstone and Crater Lake but to all national parks.

Some of my most profound and moving national park experiences have taken place when walking on ground once soaked with the blood of my fellow Americans. I recall standing next to a Civil War–vintage cannon on a warm, sultry morning in rural Maryland. Mist hung over quiet green fields punctuated by dark wood fences and scattered rows of corn. In this peaceful, idyllic landscape I could not get over what happened on September 17, 1862. On that day 23,000 Americans died during the Battle of Antietam. Antietam National Battlefield is there to remind of us of what those men and their families went through and why.

I was again reminded of that ultimate sacrifice when we visited Little Bighorn Battlefield National Monument. In southeastern Montana, Elizabeth, our son Timmy, and I walked straw-colored rolling hills under an intense summer sun. Fourteen years after the Battle of Antietam, 263 soldiers in the US Army's 7th Cavalry (including Lt. Col. George A. Custer) died on June 25 and 26, 1876, while fighting several thousand Lakota and Cheyenne warriors. According to historians, thirty-one warriors, six Native American women, and four Native American children perished on the battlefield as well. We explored the Deep Ravine walking trail, Last Stand Hill, and Indian Memorial and Custer National Cemetery. Rangers gave us vivid explanations of what happened and when. We discovered what led to the fateful day when these tribes, led by Sitting Bull, gathered for one of their last armed conflicts for justice and to defend their way of life. In the late afternoon, we headed home with a deeper understanding of what it means to be an American.

The National Park Service's 423 units take us an on a journey through our finest days as a nation. They also provide an unflinching look at our darker days. To that end, more than 500 miles east of my family's beloved Crater Lake is Minidoka National Historic Site in Idaho. It is a poignant place for my dear friend, David Nishitani, whose father's Japanese American family was forcibly moved more than 650 miles from their Seattle home to imprisonment in Minidoka during World War II.

"You wonder how they survived," he said. "No privacy. When I went there, I was struck by the guard tower—looking in, not out. What my family had to go through. Our generation has no clue." A national park ranger, Hanako Wakatsuki, helped David process what he was seeing and feeling at Minidoka.

"It's good they do that. People should understand what went on. Hanako told me they have pilgrimages to Minidoka. I plan to go next year."

Well south of Minidoka, César Chávez led a 340-mile, twenty-five-day pilgrimage from Delano, California, to Sacramento in 1966. The pilgrimage, and Chávez's resolute commitment to ameliorating the dreadful working conditions and pitiful wages of farm workers, was recognized on October 8, 2012, with the creation of César E. Chávez National Monument during Jonathan Jarvis's tenure as director of the National Park Service. In the spirit of Chávez, the pilgrimages continue today.

That is what national parks do: they inspire us to go places, meet people, think about who we are, where we came from, where we are going, and what we truly care about. And there is another benefit that is particularly important to me. National parks are a place to heal.

When we were completing the Yellowstone issue of *National Geographic* in early 2016, I was suddenly diagnosed with stage-four lung cancer. We chose to get treatment at the University of Virginia's Emily Couric Cancer Center in Charlottesville—in part because our small farm in rural Virginia was just an hour away in the Blue Ridge Mountains with a lovely view of Shenandoah National Park. Following my chemotherapy and radiation treatments, I would come home and look out the window at the park, a place where I wanted to be. Some days, when I had the strength, Elizabeth would drive me there and we would walk in the forest. We occasionally had the pleasure of bumping into park volunteers, such as Jack Price, who Elizabeth worked with as a volunteer herself. Though exhausted, I always returned home feeling better— physically and emotionally. The healing continued shortly after treatment when we visited Yellowstone to hike, camp, and thank the many people who made the May 2016 issue on the park such a success.

I returned to work at *National Geographic* headquarters and often walked a few blocks to another uplifting place, the National Mall. My love of national parks occasionally drew me to the Lincoln Memorial because it was there that President Lincoln set the stage for the birth of our park system. On June 30, 1864, as the Civil War raged, Lincoln signed legislation that gave Yosemite · Valley and the Mariposa Big Tree Grove to the State of California "upon the express conditions that the premises shall be held for public use, resort and recreation."[1] The Yosemite Land Grant was the first parkland the federal government protected for public use.[2] Lincoln knew we needed places where everyone could gather to recover from the trauma of the War between the States. Yosemite—an awe-inspiring landscape—was an ideal place for all Americans to heal.

Lincoln's idea set the precedent for another big, bold idea eight years later, the creation of the world's first national park, Yellowstone. Now we need more big, bold ideas—and the Jarvis brothers have the experience, knowledge, and wisdom to deliver them.

This book is their story, in many ways as powerful and urgent as the story the members of the 1871 Hayden Expedition told Congress and the nation after exploring Yellowstone. That is what national parks do—they inspire stories and tell stories. Unfortunately, too much of our current park story is mired in the politicization of the National Park Service, a loved and respected institution that had escaped partisanship until December 31, 1972. On that day, President Richard Nixon fired NPS Director George Hartzog, who had had a distinguished career in the Park Service. He was replaced with an inexperienced political crony. Nixon began a destructive trend that continues to undermine and haunt the Park Service. We, the people, and our parks deserve better. Jonathan and Destry clearly see the challenges and suggest thoughtful, pragmatic solutions.

Their stewardship of the bold idea that began in 1872 was well expressed by the acclaimed Western writer Wallace Stegner: "National parks are the best idea we ever had. Absolutely American, absolutely democratic, they reflect us at our best rather than our worst."

I believe this to be true. For national parks to realize their potential and evolve and become healthier, we need to work together—united in the common good for all Americans. Stegner recognized this and said, "One cannot be pessimistic about the West. This is the native home of hope. When it fully learns that cooperation, not rugged individualism, is the quality that most characterizes and preserves it, then it will have achieved itself and outlived its origins. Then it has a chance to create a society to match its scenery."[3]

Stegner's observation applies to more than the West: it applies to all of our beautiful, diverse, imperfect nation. Truly, national parks are the "best idea we ever had."

CHRIS JOHNS
National Geographic Strategic Advisor
Missoula, Montana

PREFACE

Our father, V. B. Jarvis, joined the Franklin D. Roosevelt administration's Civilian Conservation Corps (CCC) in 1935 at age twenty. He served for eighteen months in a U.S. Forest Service camp in what is today the Mount Rogers National Recreation Area in southwest Virginia. Dad was an outdoorsman in the most classic sense. He loved to hunt, fish, and roam the forests, but he also managed his own sporting goods store, complete with a hunting dog kennel. Though never formally educated, he was intelligent, curious, and a sharp observer of nature.

One of the things he taught us was to clear a small sitting area in the forest, usually at the base of a large tree. By removing any twig, leaf, or other noisemaker, you could settle in as a quiet observer. Within about twenty minutes, the forest and its residents would have forgotten your presence and out they came, to carry on their lives; we were only there to watch. Appreciating nature, whether by leaving it alone or actively managing for naturalness, has been the bedrock of our approach to the national parks for fifty years.

For the two of us, those early experiences stimulated the desire for a deeper understanding of nature, and science provided the path. Both of us graduated with degrees in biology from the College of William and Mary. While we are not scientists, we are well schooled in the scientific method, with the ability to read and understand science and to appreciate its essential importance to understanding the world in which we live. Our desire to see science applied to the stewardship of our national parks has been the centerpiece of our dual careers—Jonathan's career largely spent inside the National Park Service—from NPS ranger to NPS director—and Destry's largely spent outside as a nonprofit leader, though with a stint as a Clinton administration appointee inside the NPS as well. Over time, we both embraced and absorbed the true meaning and value of our national park system, for all Americans and others from around the world. We also saw the conflicts and political interference that at times prevented the NPS from being all that it should be. And we knew we had to stand and fight for the universal values represented by the cultural and natural national parks now found in all fifty states and the US territories.

In this book, we each tell our stories separately and then reach conclusions on each issue together. Destry's career has been primarily that of a parks advocate and policy wonk, lobbying, suing, and publicly supporting or chastising the NPS and its political controllers, as warranted. In contrast, Jonathan's career has been that of a professional ranger and field park manager as well as agency director, making challenging decisions based on science and adherence to policy, and pushing back against political appointees who do not support the mission of the National Park Service. Our overall conclusion, that the NPS should be an independent agency, has broad implications for the future of the National Park System that we hope will inspire all Americans.

DESTRY

In spring 1972, I returned to Washington, DC, from my US Army service in Vietnam. Soon after, I began my conservation career as a volunteer lobbyist for Friends of Animals, learning how to lobby Congress by pushing for enactment of the Marine Mammal Protection Act (MMPA). I was spurred to action by full-page ads in the *Washington Post* protesting the clubbing to death of baby fur seals on the Arctic ice for their pure white furs. In the course of an eight-month lobbying campaign, I had the good fortune both to see the MMPA enacted and to meet the four men who would have the greatest impact on the shape of my conservation career—Spencer Smith, a long-time public-interest lobbyist and consultant, who also happened to be a board member of the National Parks Conservation Association (NPCA); US Air Force Colonel Milton Kaufmann, a retired Defense Intelligence Agency officer who led both Friends of Animals and a marine mammal protection group, Monitor International; Stewart "Brandy" Brandborg, executive director of the Wilderness Society; and Patrick Noonan, CEO of the Nature Conservancy and shortly thereafter of the Conservation Fund, which he established with one

Secretary of the Interior Ken Salazar administers the oath of office to Jonathan Jarvis with Destry Jarvis holding the Bible. Credit: National Park Service.

of the first MacArthur Foundation "genius awards." Pat is also a long-time board member of the National Geographic Society.

Spencer was the consummate lobbyist/insider with decades of direct contact on Capitol Hill. He operated as a consultant on behalf of conservation at a time when the Internal Revenue Service tax laws did not allow nonprofit organizations to lobby Congress. His last job before retirement was as senior staff to House Speaker Tip O'Neill. He introduced me to NPCA, where I was hired as a legislative information specialist shortly after MMPA became law in December 1972.

Milt was a former military intelligence officer, as I had been, who taught me the value of coalition building so that small organizations could gain influence by joining like-minded advocates in pursuit of a common cause. Milt also invited me in the fall of 1973 to join him on the "environmental sail" on the Chesapeake Bay, a multiorganization effort to call attention to the deteriorating condition of the bay's water quality and aquatic life. I have continued to make Chesapeake Bay conservation a key aspect of my conservation career.

On that sail I also met Pat Noonan, the individual who has had the single greatest impact on my career over its entire course. When I left NPCA in 1988 after sixteen years of parks advocacy, Pat hired me to launch his Civil War Battlefields protection initiative, and he also got the National Geographic Society to hire me as chief consultant for the first edition of its *Guide Book to the National Parks*. As a board member of the Student Conservation Association (SCA), Pat was influential in the SCA's hiring me as its executive vice president in 1989. I loved the SCA, because it was the closest thing that existed to the CCC for engaging young Americans in conservation service. While I was with SCA, our largest project was the Greater Yellowstone Recovery Corps, a multiyear, multimillion-dollar, thousand-person initiative to restore and rebuild trails, bridges, and other facilities destroyed in the 1988 wildfires. After my eight years as a politically appointed assistant director of the NPS (1993–2000), I set up my consulting business, Outdoor Recreation & Park Services, LLC, and, in 2002, Pat again hired me as a consultant to support his work on Chesapeake Bay conservation and the Captain John Smith Chesapeake National Historic Trail that stretches throughout the bay and its tributaries.

In my early days of learning how political Washington works, as well as how the nongovernmental conservation organizations work together—or not—I was deeply influenced by Brandy, who convened a monthly brown-bag lunch gathering, mostly of junior staff, to distill both how success was achieved and how to organize for success. Brandy's mantra was always that "more success is possible when you don't care who gets the credit."

Student Conservation Association trail work in Yellowstone National Park after the 1988 fires. Credit: Destry Jarvis.

Spencer, Milt, Pat, and Brandy provided me with invaluable early career guidance as I entered five decades working on behalf of America's national parks.

JONATHAN

I am six years younger than Destry and, in some ways, have followed in his footsteps—same elementary and high schools, same college, same fraternity, same major in biology—with both of us finding our place in the outdoors. He came out of college during the Vietnam War and served in the military, but the war was over by the time I was out of college. Upon graduation, I outfitted a camper van and headed west with my girlfriend. For about five months, we explored the great western national parks like Glacier, Olympic, and Yellowstone. It was from this trip that the seeds of a career path were planted. Broke and broken up, I landed back in northern Virginia at my brother's house in the late winter of 1975–76. Always prone to giving assignments to his little brother, Destry handed me the environmental impact statement for the expansion of the Jackson Hole airport in Grand Teton National Park (an issue that is still standing) and asked that I read it and write up something for his work at the NPCA. It was not long after that, perhaps since I was crashing at his house and not paying rent, that I considered applying to the NPS for a seasonal position.

In March 1976, I was hired as a GS-4 Park Technician at the Bicentennial Information Center in Washington, DC, along with an extraordinarily diverse group, selected to welcome the millions of visitors expected on this 200th anniversary of the nation. Among that group I met the stunning Paula Rosenberg, who would become my lifelong partner. Over the next two years, I worked in various positions on the National Mall, including, as I like to say, a winter with Mr. Jefferson. Stationed across the windblown Tidal Basin, it was often just me and the president in his marble rotunda. Working among the monuments, greeting thousands of happy visitors, and interpreting the history of our nation got me hooked on the mission of the National Park Service. With supportive supervisors and mentors, in 1978, I landed a GS-5 Park Ranger job at Prince William Forest Park and found my footing in a natural setting.

After Paula and I married in 1980, we were off on our NPS adventure. Using the fold-out map and guide to the national parks, we circled the parks where we would hope to one day work and live, wonderful places like Yellowstone, Rocky Mountain, and Yosemite. As it turned out, we never worked or lived in a single one of those we identified! Instead we went where the opportunity presented itself. Advancement meant moving geographically, and so every three to five years we began to look for opportunities to try something new. From Prince William Forest Park, we went to Guadalupe Mountains National Park in Texas, then to Crater Lake National Park in Oregon, then to North Cascades National Park in Washington State.

I was rising in my preferred field of natural resources but maintained my "ranger skills." I landed my first superintendent position in the remote Craters of the Moon National Monument in Idaho in 1991, and from there we moved, as Paula said, "even further from civilization" to the bush of Alaska, where I was the superintendent of Wrangell-St. Elias National Park and Preserve. After five years of hauling water and winters at 40 degrees below zero, we returned to the lower forty-eight states at Mount Rainier National Park in Washington state. And in 2002, I was tapped to be the regional director of the NPS Pacific West Region, covering fifty-eight parks in the states of Washington, Oregon, Idaho, Nevada, California, and Hawaii, and the Pacific Islands of Guam, Saipan, and American Samoa. In 2009, I was called to Washington, nominated by President Obama and confirmed by the Senate to be the eighteenth director of the National Park Service.

Along the way, we raised two children, Ben and Leah, who incorporated the values they experienced in the parks into the wonderful adults they are today. Paula was the rock at home and flexible about moving, taking on new positions when they were available in these rural settings.

As I moved up from field ranger to supervisor to superintendent to regional director to director, I saw increasingly complex jobs as learning opportunities. I worked to better understand the resources in my care, the needs of fellow employees, the diverse views of our visitors, and the relationships with surrounding communities. Throughout my career, there has always been one person upon whom I could call and commiserate on park issues of any sort, no matter how complex, legal, or political. And that person is my brother, Destry. And once again, as we write this together, he lays the groundwork and I follow his lead with my stories from the field.

INTRODUCTION AND A BRIEF HISTORY OF THE NATIONAL PARKS: 1872–1972

It is high time for a fundamental change in how the National Park Service (NPS) is allowed to steward our American national park system, by moving the NPS out of the highly conflicted and overpoliticized Department of the Interior. Instead, it should be made an independent agency by act of Congress—perhaps structured like the Smithsonian Institution or the National Archives and Records Administration. Our book will make this case, built upon our more than ninety years of combined park experience.

For most Americans, the National Park Service is defined by its iconic places such as the Grand Canyon, Yosemite, and Yellowstone National Parks. But the National Park Service is far more—and its mission more complex—than often imagined. One of those who imagined the National Park Service to have a greater role in society was Dr. John Hope Franklin, renowned scholar of African American history and recipient of the Presidential Medal of Freedom.

In a 2001 advisory report to the Secretary of the Interior, Dr. Franklin said this about the NPS:

> The public looks upon national parks almost as a metaphor for America itself. But there is another image emerging here, a picture of the National Park Service as a sleeping giant—beloved and respected, yes; but perhaps too cautious, too resistant to change, too reluctant to engage the challenges that must be addressed in the 21st century.
>
> We are a species whose influence on natural systems is profound, yet the consequences of this influence remain only dimly understood. Our in-

creased numbers have altered terrestrial and marine systems, strained resources and caused extinction rates never before seen. As developed landscapes press against or surround many parks, pollutants in both the air and
water impact park resources. Our growing numbers encourage a drifting
away from knowledge about nature and our own history as a nation and a
people.

The times call for respected voices to join in confronting these issues—
voices that can educate and inspire, leading to greater self-awareness and national pride. The National Park Service should be one of these voices.[1]

The report went on to recommend the NPS embrace its mission as educa-
tor, acknowledge the connections to Native American culture, encourage the
study of the past, and assure that "no relevant chapter in the American heritage experience remains unopened."

To awaken the "sleeping giant," to use that voice freely and without censor-
ship, for it to achieve a larger mission for the American people, the National
Park Service must be freed from the current forces that make the agency captive to partisan politics and policy manipulation to suit the administration
in office at any given time. As we will describe throughout this book, for the
last fifty years the NPS has been far more subject to political controls from
high levels of the Department of the Interior (DOI) and the White House
than was the case for the first one hundred years of America's national parks.

With its roots in the military tradition, the NPS is a strongly hierarchical
agency, with each succeeding agency director adhering to the law as set forth
in the 1916 Organic Act that established the agency. The act's clear statement
is often referred to as the NPS "mission":

> The service thus established shall promote and regulate the use of the Federal
> areas known as national parks, monuments, and reservations hereinafter
> specified by such means and measures as conform to the fundamental
> purpose of the said parks, monuments, and reservations, which purpose is
> to conserve the scenery and the natural and historic objects and the wild life
> therein and to provide for the enjoyment of the same in such manner and
> by such means as will leave them unimpaired for the enjoyment of future
> generations.[2]

In carrying out the agency's mission, NPS directors may set or adjust priorities and policies as science and scholarly history dictate how the natural
and cultural resources of the national park system are to be managed, and to
respond to the nation as we change our worldviews, especially on diversifying the full history of America. The field managers, regional directors, and

park superintendents carry out policy decisions made above them, under the broad guidance from Washington and the body of laws and policy that have shaped the system, but also need both the discretion and knowledge to take independent action when issues on the ground so indicate. As we will show, the ability of the NPS director and the field managers to keep the agency on its mission, as prescribed by the 1916 Organic Act, has shifted dramatically in recent years. The judgment of professional personnel has become limited and constrained under political controls from appointed politicians.

In 1917, the first annual report to Congress from the NPS's first director, Stephen Mather, used the US Public Health Service to call attention to the deteriorated and unsafe condition of park facilities and the need for adequate funding. In those days, the NPS Director could express these concerns directly to Congress, without political repercussions. This is no longer the case today.

The popularity of the national park system has soared in the decades since World War II. Visitation currently exceeds 320 million per year and the NPS consistently ranks as one of the most popular federal agencies. Political appointees within the service's parent department, the Department of the Interior, increasingly have recognized the potential independent political power of the NPS. They have feared this power, resented it, sought to contain or diminish it, or have taken advantage of it to foster their own political goals. In all cases, the ability of the NPS director and the agency to stay on the course of conservation and historic preservation has been reduced dramatically.

Over its first half-century, 1916–1972, and under ten presidents, the NPS was managed under seven successive professional directors who ably guided the growth and management of the national park system, largely free from day-by-day political micromanagement. Their management did not diverge from the statutory mission of the agency, to preserve the parks "unimpaired for enjoyment of future generations." In those decades, it was a relatively rare thing for the Secretary of the Interior to overrule or intervene in NPS management. Nor were these directors expected to resign whenever a new administration took office. Certainly, there were far fewer political appointees throughout the department than there are today, and the NPS director had a direct path to the Secretary. By 2020, however, there were at least nine successive layers for review and approval by political appointees who stood between the NPS director's proposed policy actions, staffing needs, and budget priorities and their implementation.

Much of the time and creative energy of those seven directors in the National Park Service's first fifty years were devoted to appropriately enlarging and diversifying the composition of the system. In contrast, the service's second half-century, the period of our active engagement, has included wide and divergent policy swings, sharply contrasting between Republican and

Democratic administrations. Budget and staffing levels, the role of science to inform decisions, the composition of the system to better reflect the face of America, fights over inappropriate forms of recreation use, wildlife preservation versus hunting, and many more issues have been casualties of these swings. The energy we spent internally and externally during these "hostile" administrations was used defending the basic premise of the agency, its mission, its employees, and its budget.

President Nixon personally fired the career NPS Director George Hartzog on December 31, 1972, and appointed Ron Walker, an advance man from his reelection campaign, who became the first purely political appointee to be NPS director. From that point forward, the job of NPS director became more of a revolving door and far more politically charged. Each succeeding president chose the director to fit that administration's policy priorities and/or to stifle any dissent from within the NPS. The ability of the NPS director to speak to Congress on his/her professional views ended with George Hartzog's firing.

Over its second half-century, since 1972, the NPS has had eleven directors, five of whom did not come from within the NPS career service. In 1998 (while Destry was serving as assistant director of the NPS for Legislative Affairs), Congress enacted a new law requiring Senate confirmation of the NPS director, including a requirement that any nominee have prior qualifying park management experience. We hoped this would assure adherence to the NPS mission by future directors. Even this attempt at having professional leadership has been thwarted, most recently, and likely illegally, by President Trump, who did not appoint a director for four years. He said he liked "acting directors" because they were easier to fire.

Ever since the 1970s, with one or two notable exceptions such as President Nixon's first four years and George W. Bush's Secretary of the Interior Dirk Kempthorne, Republican presidents and Republican-majority Congresses have consistently fostered public land laws and administrative actions that have been the antithesis of sound conservation policies. They have sought federal land disposal through transfers to states or privatization through approving private mining claims to public lands; awarded long-term mineral leases at below-market prices; opposed federal land acquisitions; sold underpriced grazing permits and timber sales contracts; and have regularly opposed new natural area parks, wildlife refuges, and wilderness designations on federal lands. They have sought (but failed to achieve) authorization of a "park closing commission" to eliminate less-visited units of the system. They have wielded sharp budget cuts in presidents' annual budget requests to Congress for the NPS, which are often approved by Republican-majority Congresses. Such budget cuts, accompanied by staff reductions mandated by the White House Office of Management and Budget, then allow them to argue that the agencies cannot "take care of what they have" as further justification for opposing new additions to the system, no matter how deserving.

Over the same fifty years, however, Democratic presidents and Democrat-majority Congresses have routinely supported significant expansion of the national park system through both designation of new units and land acquisition to expand existing units; have sought new laws and policies that have improved air and water quality; and have taken numerous site-specific actions at the behest of the NPS to improve the parks. These include stopping mining in numerous parks, reintroducing wolves to Yellowstone, supporting restoration of the Everglades, seeking and winning ecological/watershed boundaries for Alaskan parks, and broadening the system to more fully represent the diversity of America.

The Democrats have not, however, been willing to fully fund the NPS at the level necessary to sustain these most important natural and cultural places in perpetuity for the enjoyment of future generations. As just one bureau within the Department of the Interior, the NPS must compete for budget against all DOI agencies, such as the Bureau of Indian Affairs or the US Fish and Wildlife Service, and must stay within limits imposed by the Office of Management and Budget (OMB). OMB gives the department a firm maximum on the amount that it can request in the budget each year, even before the department can calculate its needs. All DOI agency budget requests must fit under the OMB cap. The NPS director is on the horns of a dilemma, advocating for increases in park budgets knowing it will mean less for Native American schools or the protection of endangered species.

One of the most egregious examples of overly politicizing the NPS occurred during the four years under President Trump. His DOI appointees feared the agency's popularity and influence, and rejected the NPS preservation mission. As a means to circumvent the Senate requirements, and to tightly control NPS policy, the president and DOI Secretaries Ryan Zinke and David Bernhardt caused the NPS to function (poorly) without a Senate-confirmed director for the only time in its history. Two of the four "acting" directors, both long-time political appointees who could not meet the statutory qualification requirements, controlled the agency at different times in Trump's four-year term. The Trump appointees knew that the NPS had too much public support for them to outright dismantle the agency, but they could keep it leaderless and adrift.

The places and stories of America in the care of the National Park Service are too important to the nation, to our high ideals, and to future generations to be subject to the whipsaw of Washington's politics. We've chosen to make our case using a unique approach that combines our personal experiences with policy analysis. The book is both a double memoir, which (we hope) gives the story life, but also a serious proposal for reform, which we believe is absolutely necessary if the national park system is to thrive in an uncertain future.

Chapters 1 through 6 each chronicle our separate but often parallel experiences and actions focused on the NPS and adherence to its mission over the last five decades. Chapter 1 covers the growth of the system and various fights with administrations and Congresses to enable the service to actually encompass its intended inclusion of the best natural and cultural places of America's heritage, fully representing the diversity of America and its history. Chapter 2 focuses entirely on the national parks in Alaska, now over half of the 85 million acres managed by the NPS, how they came about, and how they are managed differently. Chapter 3 chronicles the troubled odyssey of the NPS *Management Policies*, which sounds very bureaucratic but actually has been the most politically charged and turbulent element of NPS history over these five decades. Chapter 4 covers the application, or lack, of usable knowledge derived from scientific and historical research to the management of the national parks. Chapter 5 describes the ever-expanding role that nonprofit partner organizations can play in assisting the NPS in its mission, when their role is not used to reduce or eliminate the career NPS employees. Chapter 6 summarizes numerous examples of political interference with the NPS's ability to achieve its intended mission, "to preserve the parks unimpaired for enjoyment by future generations." Chapter 7 presents our final analysis and recommendations for the best way whereby the NPS ought to be freed from so much political interference: by reauthorizing the National Park Service as an independent agency.

A Brief History of the National Park Service, 1872–1972

In order to fully understand the NPS of today and its overpoliticized operations, we here briefly describe the first hundred years of the US national parks, 1872–1972, when the NPS was comparatively free from partisan politics that sought to reshape the agency mission and priorities. This history contextualizes the substantive content of this book, which is focused on the last fifty years. The two of us have directly experienced this period, during which time the NPS has been overtly controlled by partisan politics that conflict with its own mission.

Following the establishment of Yellowstone National Park in 1872, the composition of the US national park system evolved by acts of Congress and proclamations of national monuments by most presidents. These sites were typically under the protection of the US Army, and many of them were patrolled by the famous Buffalo Soldiers. When the National Park Service was established by act of Congress in 1916 as an agency in the Department of the Interior, there were already thirty-eight designated national parks and national monuments, all in the western states, which were put under NPS management at that time. Moving forward, both new park proposals and uniform management policies stemmed from the professional recommendations of NPS senior leaders in direct negotiations with members of Congress. By 2021, the system was composed of more than 400 units.

While the national park system was growing in scope and popularity, the NPS was also seeking to professionalize park management. At the time, there were no scientists at the Department of the Interior involved in national park management. The earliest scientific studies of national parks, such as the 1918 *Wild Animals of Glacier National Park* by Vernon and Florence Merriam Bailey, were basic field science, with no reference to policies or how the park should be managed. In 1927, the NPS hired its first scientist, George Melendez Wright, tasking him to prepare the scientific studies of park wildlife known as the *Fauna Series*. Wright's pioneering work suggested that the NPS recognize the importance of science-based conservation decision-making as well as respect for indigenous knowledge.

The emerging "system" of parks continued to expand. The first NPS unit in the East was Sieur de Monts National Monument (today Acadia National Park), proclaimed in 1916 from donated lands, six weeks before the Organic Act became law. Congress and presidents, with the leadership and initiative of the NPS and its first directors, added another thirty-two national parks and national monuments to NPS management between 1916 and 1933. At that time, most of the seventy NPS sites were natural areas, along with a few Native American archeological sites. Nearly all of these were sites that either NPS Director Stephen Mather or his successor, Horace Albright, specifically sought to have added to the growing system.

Both Mather and Albright, with little or no interference from the DOI Secretary, led a quiet campaign to get Congress and the presidents to include preservation of American history sites as a function of the NPS. By 1933, Director Albright managed to get three historic sites added to the system by individual acts of Congress— Colonial National Historical Park (Jamestown and Yorktown), Morristown National Historical Park (Revolutionary war winter encampment), and George Washington Birthplace National Monument.

The composition of the emerging national system of parks changed dramatically in 1933 when President Franklin Delano Roosevelt signed Executive Order 6166, which instituted the most significant shift in the NPS mission up to the present day. The story goes that Director Albright rode by car with FDR to visit the Civilian Conservation Corps (CCC) camp in Shenandoah National Park and seized this rare opportunity to urge the president to put the NPS in charge of the nation's most important historic sites.

With the FDR order, the NPS became the keeper of the most important places in American history, including all the battlefields then managed by the War Department as well as twenty-one national monuments that had been proclaimed by various presidents under the 1906 Antiquities Act. In addition, all "National Capital Parks," including the monuments on the National Mall, were also transferred to the NPS. Overall, FDR's action transferred twelve natural areas and forty-four historic sites to NPS management.

One of FDR's many great civic actions was to establish the Civilian Conservation Corps in 1932, which provided some 3 million young men with employment and job skills working on public conservation projects, including in national parks. NPS Directors Albright and Arno Cammerer played critical roles in setting up and managing some seventy CCC camps in the national parks. The CCC built much of the infrastructure in these national parks, much of it still in use today.

Despite the impact of the Great Depression years on the national economy, Congress and Roosevelt continued to add sites to be managed by the NPS, including numerous historic sites as well as the Everglades (1934), Big Bend (1935), Joshua Tree (1936), and Organ Pipe Cactus and Capitol Reef (1937).

One unanticipated outcome of the depression was that the NPS was given numerous public recreation sites to restore and manage, including recreation lands around large reservoirs, public parkways, and the first national seashore at Cape Hatteras in North Carolina. As the depression ended, the NPS retained three CCC-rehabilitated recreation sites near Washington, DC, that are today part of the national park system—Catoctin Mountain Park, Prince William Forest Park, and Greenbelt Park. Many other such CCC-rehabilitated sites became state parks.

In 1941, the NPS released the first of a series of site studies, authorized under the Park, Parkway, and Recreation Area Study Act of 1936, to identify places that ought to be permanently protected, either by the NPS or as state or local parks. "A Study of the Park and Recreation Problem in the United States" focused broadly on the need for public spaces for outdoor recreation. Understandably, action on many of the report's recommendations was put on hold until after the end of World War II. During the war, the entire national park system was closed to the public, and some members of Congress, along with industry allies, made repeated calls to open the parks to timber cutting, cattle grazing, and mining to support the war effort. In order to protect the parks from these unnecessary conversions, NPS Director Newton Drury effectively argued to the American people that the values represented by the parks were what we were fighting for, and, more practically, he ordered the park lodges and hotels opened for rehabilitation of returned soldiers. For example, the Ahwahnee Hotel in Yosemite became the US Naval Special Hospital for the duration of the war.

Despite the depression and war, between 1934 and 1951 another fifty-nine sites were put under NPS management. Public demand for outdoor recreation then spiked all across America in the 1950s, even while the NPS was still reeling from being shut down and understaffed during the war. Visits to the national parks rose from 6 million per year in 1942—even though the parks were closed to the public at that point—to 72 million by 1960. The popularity of the parks combined with their deplorable conditions led NPS Director Connie Wirth to meet directly with and so convince President Eisenhower to launch "Mission 66" in 1956 (with completion timed to recognize the fiftieth anniversary of the NPS in 1966). Mission 66 was a decade-long plan to restore the parks and modernize facilities. Included in the plan were resumption of the service's National Survey of Historic Buildings and Sites and the launch of a parallel process to identify nationally significant natural areas to be designated as national natural landmarks. These surveys led to many subsequent additions to the system.

Although the idea that the NPS could manage the national parks without adequate knowledge, derived from good science and applied research, persisted in these early years, it began to change rapidly in the 1960s. An initial push came as a result of the First World Conference on National Parks, held in Seattle in 1962. A fifteen-member, eight-nation committee at that conference produced a report, "Management of National Parks and Equivalent Areas," which concluded that "few of the world's parks are large enough to be in fact self-regulatory ecological units; rather most are ecological islands subject to direct or indirect modification by activities and conditions in the surrounding areas." It went on to state that "management based on scientific research is, therefore, not only desirable but often essential to maintain some biotic communities in accordance with the conservation of a national park."

The next significant event to dramatically change the NPS was the publication of the January 1962 Outdoor Recreation Resources Review Commission's (ORRRC) report to Congress. ORRRC was initially appointed by President Eisenhower and chaired by Laurance Rockefeller; it was established to make recommendations for the future of outdoor recreation in America. The ORRRC report, submitted to President Kennedy, broadly set five major recommendations for improvement of outdoor recreation:

- Setting a national outdoor recreation policy
- Developing guidelines for management of outdoor recreation resources
- Expansion, modification, and intensification of present programs to meet increasing needs

- Establishment of a Bureau of Outdoor Recreation (BOR)
- Establishing a federal grant-in-aid program to states[3]

The impact of the ORRRC report and its numerous specific recommendations, coupled with the strong support of Presidents Kennedy and Johnson and Interior Secretary Stewart Udall, who served through both administrations, was enormous. These federal laws owe their origins to the ORRRC report: the Outdoor Recreation Act, The Wilderness Act, the Land and Water Conservation Act, the National Wild and Scenic Rivers Act, the National Trails System Act, the Youth Conservation Corps Act, and the Water Projects Recreation Act, all of which were enacted by Congress between 1964 and 1970. Each of these laws has also had a major effect on the National Park Service, both as to composition of the system and use of science-based management policies.

Unique among these laws, the 1964 Land and Water Conservation Fund (LWCF) for the first time provided direct funding for federal land acquisition for national parks and other public land agencies by appropriating revenues from oil and gas leasing on the outer continental shelf. The rationale is that, as one finite publicly owned resource is permanently depleted, some of the revenue ought to be dedicated to acquiring another permanent public resource, in this case parks, refuges, and other designated areas. Virtually every new national park system unit established since then has used LWCF funds to make the park a reality. Coupling this new acquisition funding with the still-increasing demand for outdoor recreation space, over the next decade Congress authorized ten new national seashores and lakeshores, five urban national recreation areas, and nine national rivers and flat-water recreation areas. Authorized at $900 million annually, Congress has historically appropriated only a portion to the states and federal land management agencies.

Up to that time, the NPS had carried out the lead function of nationwide outdoor recreation planning, but its Office of Nationwide Planning and Co-operative Services was transferred to the newly established Bureau of Outdoor Recreation in 1963 upon enactment of the National Outdoor Recreation Act. The NPS resented this change and loss of function. Its first response to these changes was to separately publish *Parks for America* in 1964, a large format, heavily illustrated, 500-page publication that covered all fifty states, recommending site-specific places for future designation as local, state, and federal park lands. In those days, the NPS could do that, but in recent administrations, the agency would never be allowed to take its own position on much of anything, much less publicly release a competing policy document.

As noted in DOI Secretary Stewart Udall's foreword to *Parks for America*, "at the time the Bureau [of Outdoor Recreation] was established, the National

Park Service, in cooperation with the States, had just completed a draft of this volume consisting of a report on existing non-urban park and related recreation areas and on approximately 2,800 potential areas. The proposals in the report must be regarded as the views of the National Park Service. They do not necessarily reflect the views of the Bureau of Outdoor Recreation." For the NPS, the report included proposals for forty-two future units of the national park system. Forty of those sites have since been made units of the system by acts of Congress.

Continuing a national recognition of the need for more outdoor recreation opportunities, but with a new focus on the impacts of industrialization on the natural environment, President Lyndon Johnson and First Lady "Lady Bird" Johnson convened a White House Conference on Natural Beauty in May 1965. In 1966, it was followed by the appointment of the President's Council on Recreation and Natural Beauty, chaired by Vice President Hubert Humphrey, and a parallel Citizen's Advisory Board on Recreation and Natural Beauty, chaired (again) by Laurance Rockefeller. The President's Council rendered its report in 1968, entitled "From Sea to Shining Sea: A Report on the American Environment—Our Natural Heritage." BOR Director Ed Crafts served as executive director of the council for its duration.

Conservative policy pushback against these conservation and recreation policies came in 1964 when Congress authorized P.L. 88-606, the "Public Land Law Review Commission" (PLLRC), an initiative of House Interior Committee Chair Wayne Aspinall (who was also the self-appointed chair of the commission). Aspinall envisioned the PLLRC report as a direct counter to the Kennedy/Johnson proconservation, new park, and outdoor recreation public lands initiatives.

The product of the PLLRC's work was its June 1970 report entitled "One-Third of the Nation's Land: A Report to the President and to the Congress."[4] The report placed heavy emphasis on the need for development of public resources through mining, timber harvest, and livestock grazing, with a particular focus on continuation of the long-standing desire to "dispose" of more federal lands by transfer to the states or sale for private ownership. This initiative was also meant in part to forestall the growing conservation movement's successes in having more federal lands designated by Congress as parks and wildlife refuges.

Directly pertinent to the main thesis of our book—the need for greater professional autonomy and independence for the NPS versus the impact of partisan politics on the agency—the two terms of President Nixon stand in sharp contrast to each other. In his first term, 1969–72, Nixon visibly embraced environmental protection and land conservation as a national priority. More specifically he supported major new laws for clear air, clean water, and national land-use planning; his "Parks to the People" initiative supported urban national recreation areas, including Gateway National Recreation Area (NRA) in New York City and Golden Gate NRA in San Francisco, which turned largely surplus federal lands into public parks for urban dwellers.

The decade from 1962 to 1972 was progressive in the best sense of that term for the NPS, with growth in units, staff, and budgets. But it also saw new conflicts arise in reaction to the growing popularity of the NPS preservation mission, as amply demonstrated by the two official reports cited above with opposite conclusions on public land conservation. The ORRRC report to Congress in 1962 and the Congressional report "One-Third of the Nation's Land" in 1970 exemplify the diametrically opposite policies—public land conservation and outdoor recreation versus mining, timber cutting, and grazing—that emerged in the 1970s and continue up to the present, swinging between Republican and Democratic administrations and Congresses. Both of these reports, however, essentially ignored the role of the NPS which, up to that time, had been allowed to pursue its own agenda. This greater professional freedom came to an abrupt halt with Nixon's firing of NPS Director Hartzog at the end of 1972.

In anticipation of rising political concerns over growth of the national park system, in June 1970 the Secretary's Advisory Board on National Parks, Historic Sites, Buildings and Monuments approved a new policy on expanding the system. Carefully engineered by NPS Director George Hartzog, the policy stated that it was establishing "valuable guidelines for further evolution of the National Park System Plan and a useful framework within which to present plans and priorities . . . for expansion of the National Park System."[5] As a result of this approval, in 1972 the service published two volumes—*Part One of the National Park System Plan: History*, and *Part Two of the National Park System Plan: Natural History*. These two volumes taken together, called the *National Park System Plan*, became the justification for Director Hartzog to lead a national effort to further expand the system. Despite growing opposition from other federal agencies, the Nixon administration, and industries seeking resource development and extraction, Hartzog succeeded in adding most of the proposed sites during his tenure.

Those specific sites identified in the *National Park System Plan*, along with the idea that the NPS should have a professional leadership role in decisions about the future composition of the system, have guided much of our work on behalf of our nation's national parks since their publication in 1972. Hartzog's continued public support for adding new areas to the system contributed to his untimely and ill-advised firing in December 1972.

Over that first century of the national parks, the succession of Congresses, whether Democratic or Republican majority, generally kept to their principal functions as they apply to the NPS: authorizing new parks and appropriating funds to manage them. These congresses showed great deference to NPS recommendations on whether or not a proposed site was qualified for addition to the system. Critical "oversight" of the NPS management by Congress was a rarity.

Two significant government reforms initiated by the Nixon administration in 1970 have had a profoundly negative impact to this day on how the NPS, and most other federal agencies, do their jobs. First, President Nixon expanded White House control of the federal bureaucracy by changing the White House Bureau of the Budget into the Office of Management and Budget (OMB). Previously, presidents set broad policy and priorities by how the federal budget was allocated but had little to do with monitoring or managing day-to-day regulations or programs of the various agencies. By asserting responsibility for coordinating and controlling "management" in all federal agencies, OMB, which has clearly become the most powerful entity in the federal government, is able to assure that the president's priorities and preferences are carried out at all levels of the federal establishment. OMB not only controls the budget request to Congress for each agency annually and in five-year planning cycles, but it also sets the number of employees each agency may hire every year and must approve every new regulation proposed by any agency.

Second, prior to the Nixon administration, politically appointed and Senate-confirmed assistant secretaries, such as the assistant secretary for Fish and Wildlife and Parks in the DOI, were merely senior staff to the Secretary and had no line authority over the bureaus assigned to them. For example, the NPS director reported directly to the Secretary under the old system and did not take orders from or report to the assistant secretary for Fish and Wildlife and Parks.[6] Today, based on the Nixon administration's decision to increase the power of assistant secretaries, and thus political control, every major agency decision, including selecting park superintendents, must be reviewed and signed off on at the assistant secretary level, prior to any contact with the Secretary. In fact, any NPS initiative or decision must now also be approved by ten people above the NPS director: two deputy assistant secretaries for Fish and Wildlife and Parks, the assistant secretary for Fish and Wildlife and Parks, a deputy assistant secretary for Policy, Management and Budget, the assistant secretary for Policy, Management and Budget, the associate solicitor for Conservation and Wildlife, the solicitor, the deputy secretary, the secretary's chief of staff, and finally the Secretary. The NPS cannot possibly preserve, unimpaired, the varied natural, cultural, and recreation areas of the national park system for the benefit and enjoyment of future generations while being overruled, second-guessed, threatened, and/or ignored by purely partisan political appointees who control its every decision, personnel change, and budget request.

ONE

GROWING THE SYSTEM AND TELLING A MORE COMPLETE STORY

Since Yellowstone National Park was established in 1872, the national park system has grown to over 400 national parks, monuments, recreation areas, battlefields, and historic sites. Documentary filmmaker Ken Burns called it "the Declaration of Independence written on the ground." Wallace Stegner said, "The National Parks are the best idea we ever had. Absolutely American, absolutely democratic, they reflect us at our best rather than our worst." In the establishment of each park unit by an act of Congress or presidential proclamation, there is an intent to preserve a special place and its natural and cultural history for present and future generations to experience. Like a professional curator of a great museum, the National Park Service has tried to guide this growth with science and scholarly research so that the system represents, accurately and completely, the best of our natural landscapes and the most important sites of our history, including the periods when we as a nation failed to live up to our ideals. Unlike the broad bipartisan support that the NPS enjoyed from Congresses and presidents for its first fifty years, in the last half-century NPS growth has been subjected to the whipsaw of political appointees, presidents, and members of Congress who have advanced, stopped, resisted, rushed, and sometimes attempted to reduce the number of places in the national park system. In addition, this haphazard growth, driven by political agendas, has left the system chronically underfunded. The two of us have been in the middle of these challenges for the last forty-eight years, working inside and outside to manage and expand the system so that it represents the very best of our nation, to be preserved in perpetuity.

DESTRY

As President Nixon's second term commenced in 1973, I had only recently returned from US Army service in Vietnam and barely begun my career focused on the national park system and National Park Service. Nixon's 1974 resignation and shortened tenure were not only highlighted by the Vietnam War and Watergate but by a remarkable turnaround in his conservation and environmental policies. He switched to a much more conservative set of policies: for example, suddenly opposing new additions to the national park system, especially urban units, after having supported them in his first term.

My first and only meeting with George Hartzog while he was NPS director (I met with him many times in the years after) took place at the office of the National Parks Conservation Association (NPCA) in Washington, DC. It was early December 1972, just two weeks after I was hired by NPCA, and, as it developed, just two weeks before Hartzog would be summarily fired by President Nixon. Hartzog had come to meet with us to discuss his National Park System Plan, seeking NPCA's support. Senior staff in the NPS told me the reason for his dismissal immediately thereafter was that Secretary of the Interior Wally Hickel had tired of Hartzog's constant advocacy for new additions to the system, especially for the new (and very expensive) idea of urban national recreation areas. The story of Hartzog's dismissal, told to me directly by a longtime White House staffer, was that one day in the White House, Nixon put his arm on the shoulder of his Florida crony, Bebe Rebozo, and said "Bebe, you have done so much for me, what can I do for you?" To which Rebozo replied, "Fire that son-of-a-bitch who runs the Park Service!"—which the president ordered shortly thereafter. As that story unfolded, Rebozo had gotten a ticket from a park ranger in Biscayne National Park for tying his boat illegally to an NPS administrative dock there. Rebozo sought to have it canceled up the NPS chain of command, to no avail.

To double down on this first-ever, but far from last, political assault on the NPS directorship, President Nixon then appointed one of his campaign advance men, Ron Walker, to the job. Walker was an unqualified appointment, openly admitting that he did not know the difference between the National Park Service and the Boy Scouts. At his first public hearing before the House Subcommittee on National Parks, chaired by Roy Taylor (D-NC) and which I attended, Walker was asked a question about Yosemite National Park. Walker immediately turned to his staff assistant, and in a voice heard throughout the chamber, asked "What state is that in?" Once the laughter and embarrassment subsided, the hearing went sharply downhill from there. Walker's tenure as director was fraught with controversy, and, along with the much larger controversy that became the Watergate scandal, during which time many political appointees resigned early rather than get caught up in the investigation of illegal activities, he resigned the job after barely two years. After this, the job of NPS director became more of a revolving door; far more politically charged, the job turned over as each succeeding president selected a director to fit that administration's priorities.

Congressional initiatives had established the first two urban national recreation areas earlier in 1972, with Hartzog's strong support—Gateway NRA in New York City and Golden Gate NRA in and around San Francisco. The core of each of these new parks was surplus federal land that was transferred to the NPS by the legislation. However, the high cost of restoring, maintaining, and adequately staffing these heavily visited parks soon was apparent to the Nixon administration, and it quickly sought to label these "demonstration areas," publicly opposing the addition of any other urban NRAs to the NPS. Nevertheless, and over strenuous resistance from the Nixon administration, Congress added the Cuyahoga Valley NRA in Ohio to the system in 1974.

In early 1973, the Nixon White House had ordered the Bureau of Outdoor Recreation's (BOR) *First Nationwide Outdoor Recreation Plan* to be replaced by a much revised and slimmed-down version, which was released in December 1973. Entitled *A Legacy for America*, it offered the view that everything that the federal government needed to do for outdoor recreation had been done, and instead it emphasized actions that ought to be taken by state and local governments. For the NPS, the Nixon administration began to talk about "rounding out the system," as if the national park system were nearly completed. As NPCA's legislative director then, I wrote to Congress and testified at hearings about how this policy was wrong-headed: it ignored the clear gaps in the composition of the national park system, which still lacked numerous historic sites, seashore and lakeshore recreation sites, and places that would better diversify the system to represent all of America's natural and cultural history.

By 1974 the Nixon administration had continued to refuse to publicly release the BOR's internal draft of the *First Nationwide Outdoor Recreation Plan*, because it was thought too expensive and put too much emphasis on federally managed conservation and outdoor recreation lands and programs rather than other (extractive) uses of public lands. As a consequence of this refusal, in a very unusual move, the chair of the Senate Committee on Interior and Insular Affairs, Henry M. "Scoop" Jackson, arranged for the original BOR plan to be published as a congressional document, entitled *The Recreation Imperative: A Draft of the Nationwide Outdoor Recreation Plan Prepared by the Department of the Interior.*[1] While the White House–approved *Legacy for America* put emphasis on state and local actions, the far more detailed *Recreation Imperative* sought to put the focus on actions needed by the other federal land management agencies—the Bureau of Land Management (BLM), US Fish and Wildlife Service (USFWS), and US Forest Service (USFS)—rather than the NPS.

From the very beginning of my national park work, it was very clear that there was a major disconnect between the expansion of the system by addition of new units to manage and protect and then securing their adequate funding and staffing. It was far easier to lobby the authorizing committees in Congress to add a new nationally significant site to the national park system than it was to get the administration to request the needed funds or for the House and Senate Appropriations Committees to appropriate federal funds to manage the system at the level needed. I and my NPCA staff lobbied Congress for both annually, but the two processes have continued to diverge over the ensuing decades, resulting in serious underfunding of park operations, staffing levels, resources management, and facilities maintenance.

Over the course of 1975 and 1976, I worked closely with staff attorney Ron Tipton of the House Subcommittee on Environment, Energy, and Natural Resources on gathering data for a major report issued by the full House Government Operations Committee in June 1976, entitled simply *The Degradation of Our National Parks.*[2] Major findings included that "the Office of Management and Budget is primarily responsible for the inadequate resources that have been allocated to the Park Service to protect and operate the national park system" and that "the acquisition and development of new park areas and the construction of new facilities in the National Parks without a corresponding increase in the NPS budget for park operations and maintenance has contributed to the Park Service's inability to effectively manage park resources." Without a doubt, one of the single most important changes needed in order for the NPS to operate professionally at the level Congress has authorized was recommended in this report: "The Park Service should be allowed the freedom to develop budget requests based solely on its professional judgment concerning the needs of the National Park System, and without regard to the limitations presently imposed by OMB." This capacity is still lacking today.

Since Gerald Ford had been a park ranger in Yellowstone NP during college, I had hoped that he would be sympathetic to the NPS mission. But his two years as president (1975–76) included three different Secretaries of the Interior, one of whom, Thomas Kleppe, who had previously been Secretary of Agriculture, brought a US Forest Service–induced disdain for the NPS to the job. There were no NPS-related initiatives fostered by the Ford administration until his election campaign began. Then Ford returned to Yellowstone to announce his "Bicentennial Land Heritage Program," which proposed no new parks though it did intend to recommend new maintenance funding for the NPS budget, if he was elected.

I spent a considerable amount of time for NPCA in 1976 lobbying Congress to add the Congaree Swamp to the national park system. This last remaining intact black-water, river-bottom hardwood swamp, which had been privately owned by a family since before the Civil War, contained a remarkable number of national record–sized trees of various species. Although some politicians and political appointees scoffed at the idea of adding a swamp to the system, arguing "who would ever visit a swamp," they failed to appreciate that parks are not only about visitor numbers but about preserving the best of the best. Congaree became a national monument in October 1976 and was redesignated as a national park in 2003.

Jimmy Carter took office as president in January 1977, essentially at the same time as Congressman Phillip Burton became chair of the House National Park Subcommittee. During the four-year term of President Carter and Burton's six years as subcommittee chair, I worked more closely than ever with both NPS professional staff and Carter's DOI appointees, as well as with Chair Burton's subcommittee lead staffer Cleve Pinnix and Burton himself on numerous park bills, beginning with the Redwood National Park Expansion Act.

In late summer of 1976, I traveled to San Francisco for a first-hand view of the heavily eroded, clear-cut lands around the park and to testify at a congressional field hearing about the damage being done to the park from the clear-cutting around it. Burton's California colleague Leo Ryan, chair of the Subcommittee on Environment, Energy, and Natural Resources of the Committee on Government Operations, held the hearing.

Phillip Burton was without doubt the most skilled legislative tactician with whom I have ever worked.[3] Shortly after he assumed the subcommittee chair in early 1977, Sierra Club president Dr. Edgar Wayburn, a close Burton ally from his San Francisco congressional district,[4] came to DC and convened a small meeting of park advocates, including myself. Wayburn described Burton as "a big train engine barreling ahead, and the job for the rest of us is to lay track in front of him in the direction we want him to go."

Burton's first legislative move in early 1977 was to push expansion of Redwood National Park, which had been established in 1968. The goal was to put a stop to rampant clear-cutting of old-growth coast redwood trees on private lands that surrounded that original park. Over that winter, I worked closely with Ryan's subcommittee staff attorney Ron Tipton as he prepared the subcommittee's report on the plight of the redwoods, *Protecting Redwood National Park*,[5] which was approved by the full committee and issued

Clear-cut scar above Redwood National Park boundary. Credit: Destry Jarvis.

in March 1977. The report called for a stop to clear-cutting and for a major expansion of the park by purchase of the private timber lands upslope from the existing park boundary.

While Burton's Redwood Expansion Act was pending before Congress, the private timber companies that owned these virgin redwood trees would not halt clear-cut logging of their lands that were proposed to be added to the park, as Burton had requested them to do voluntarily. Consequently, he added a unique provision to the bill, which I testified in full support of, termed "a legislative taking"—which meant that ownership of all the lands being added to the park transferred instantly to federal ownership when the president signed the act into law, with payment to be determined later through the federal courts. The Redwood Expansion Act became law on March 27, 1978.

Even at this early stage of his time as subcommittee chair, while he was mostly focused on his home state of California, I worked with Burton and his professional staff to quietly slip a major clarifying amendment to the NPS General Authorities Act of 1970 into the Redwood bill. Known as the "nonderogation" standard of park management, this important amendment states that the "authorization of activities shall be construed and the protection, management, and administration of these areas shall be conducted in light of the high public value and integrity of the National Park System and shall not be exercised in derogation of the values and purposes for which these various areas have been established, except as may have been or shall be directed and specifically provided by Congress."[6]

After the Redwood Expansion Act, Burton shifted his approach from a single-subject bill to a multisubject bill, which I worked on extensively with his staff for the next six months. Cleve Pinnix and Clay Peters, his Republican counterpart on the subcommittee staff, were both former NPS park rangers with deep knowledge of how parks work and what legislative authorities would work best. I worked with them weekly for nearly two years as they put together what became the largest national parks bill ever assembled, before or since. At the same time, I was working with their Senate committee staff counterparts, Tom Williams (D) and Tony Bevinetto (R), on essentially the same legislative proposals. For twenty years, from the mid-1970s to the mid-1990s, I played handball and racquetball at 7 a.m. three days a week with Tom Williams. He was one of my best friends and died far too young.

For this new effort, Burton's key tactical maneuver was to learn from the regular successes of the Public Works Committee to put together a multipart "omnibus" bill that included a diversity of bipartisan components in order to engender broad support for the whole bill. Such an omnibus bill had never been done for parks legislation, but Burton reached out to all political divisions of the House, adding provisions into the package so that virtually every member had a pet project included. The bill became the National Parks and Recreation Act of 1978, Public Law 95-625 (referred to hereafter as the 1978 Omnibus Act), passing the House 341–60.[7]

The bill established eleven new national parks, historic sites, and national seashores, authorized additional segments of eight rivers to the Wild and Scenic River system, and designated four new national trails. Studies for future inclusion in the Wild and Scenic River system would be authorized for another eighteen rivers.

In an exclusive three-hour interview excerpted in our May 1979 NPCA magazine, Burton said the bipartisan omnibus bill was built on trust. He told *National Parks and Conservation Magazine* editor Joan Moody:

> You'd be shocked at how far my colleagues will reach if they know that the situation is all on the level . . . that you're not going to demagogue them from the left or the right or up or down—so that when they talk to their constituencies . . . they're assured they won't be badmouthed by the opposite constituency.
>
> By contrast, one reason that I am a total enemy of the oil and gas industry is that I don't trust them at all—I can't tell where those bastards are going to end up. Environmentalists have to terrorize the bastards, you have to

> *To Destry Jarvis, Thanks for your advice and your help in making this dream come true. Congressman Phil Burton*

95th Congress
2d Session

COMMITTEE PRINT NO. 11

LEGISLATIVE HISTORY OF THE
NATIONAL PARKS AND RECREATION ACT OF 1978
(PUBLIC LAW 95–625)

COMPILED BY

SUBCOMMITTEE ON NATIONAL PARKS AND
INSULAR AFFAIRS

OF THE

COMMITTEE ON INTERIOR AND INSULAR AFFAIRS

OF THE

U.S. HOUSE OF REPRESENTATIVES
NINETY-FIFTH CONGRESS
SECOND SESSION

DECEMBER 1978

Printed for the use of the Committee on Interior and Insular Affairs

Congressman Phillip Burton's inscription to Destry on Omnibus Bill Report cover. Credit: Destry Jarvis, personal copy.

learn to terrorize them. I made reference to that when I said [in the case of the redwoods] you have to fight for fundamental policy and then for add-ons that are peripherally expendable.

Another generic provision of the omnibus bill was a statutory requirement for the NPS to study and evaluate sites all across the country for possible addition to the national park system, and to submit a list of at least twelve such qualified sites annually to Congress. Both the NPS and I actively supported this idea, since the agency's professional judgment of a site's national significance ought to be a deciding factor in what gets added to the system.

However, in one meeting in which I voiced strong support for the importance of the NPS's professional role in determining national significance, Chair Burton responded, "NPS should stick to doing the studies as we require, and the Congress will tell the American people what is nationally significant by what we enact."

Although there were many provisions of the 1978 Omnibus bill that I and other NPCA staff worked to support, several stand out as especially important in pushing the NPS in new directions. New NPS urban units were authorized at Santa Monica Mountains NRA in the Los Angeles metro area, the Chattahoochee NRA in and around Atlanta, and Jean Lafitte National Historical Park in and around New Orleans. These were all created with strong support on the Hill from myself on behalf of the NPCA and from the Carter administration.

More broadly, the 1978 Omnibus bill authorized the Urban Park and Recreation Recovery Program (UPARR), a competitive grant program for the NPS to provide needed funds for parks in the larger US cities to support both facility maintenance and recreation programming. These funds were desperately needed, since the state grant program of LWCF largely bypassed funding for city parks. My staff and I strongly advocated for UPARR, along with the Carter administration. In the Department of the Interior, the newly established (and short-lived) Heritage Conservation and Recreation Service (HCRS) was an especially strong supporter. An expanded and renamed Bureau of Outdoor Recreation, HCRS had a decided urban focus in program interests. HCRS was the brainchild of Carter appointee Chris Delaporte, its director. In 1977, he had the history and cultural programs of the NPS, including the National Register of Historic Places, transferred out of the NPS and into HCRS/BOR. The NPS fought against these moves until eventually HCRS/BOR was abolished entirely and its program and staff folded back into the NPS in 1981 by Interior Secretary James Watt, who had been BOR director in the Nixon administration.

Undoubtedly the most innovative provision of the 1978 Omnibus bill was authorization of the 1-million-acre Pinelands National Reserve in New Jersey, which was the first federal attempt to engage in large landscape conservation by means other than federal ownership and direct management. The basic concept, known as "greenline parks," was roughly modeled after the British national parks, where very little land is publicly owned but for which the public has certain rights of access for recreation. In the United States, the closest model is Adirondack State Park in New York. I worked closely on this provision of the bill with the Pinelands Preservation Alliance, which was then and is still today well led by its Executive Director Carleton Montgomery. The act incentivized the State of New Jersey to authorize a commission with broad land-use planning authority, superseding that of local counties within the 1-million-acre boundary (the green line). The Omnibus bill required the commission to develop a comprehensive land use plan and submit it to the Secretary of the Interior for approval, which would trigger release of federal funds to support implementation of the plan. The NPS was provided with one seat on the commission board by statute.

The NPS was authorized to fund 75 percent of the cost of the plan as well as 75 percent of the cost of all land acquisitions under the plan (LWCF annual state assistance grants only cover 50 percent). The provision of New Jersey state law authorized for the plan that has made it such a huge success over its more than forty-year history is the authority—known as transfer of development rights (TDR)—that requires urban developers to acquire conservation easements over rural forest and farm lands in return for increased density for developments in small cities and towns within the boundary of the reserve. Today, some 600,000 acres of the reserve are protected in perpetuity through such TDR easements.

Another part of the Omnibus bill that I worked hard on authorized the New River Gorge National River in West Virginia, added by Burton at my request as he engaged in "shuttle diplomacy" back and forth between House and Senate. I had already spent nearly two years in and out of West Virginia, working closely with my good friend Jon Dragan, founder/owner of Wildwater Expeditions. Together we negotiated all the provisions of the bill with numerous local businesses, citizens, and elected officials, including writing the text of the bill that became law, at the behest of both House and Senate members of the West Virginia delegation. The day the bill was being considered in the Senate, I met again with West Virginia Senator Jennings Randolph, chair of the Public Works Committee. He promptly called the Army Corps of Engineers district engineer in West Virginia and said, "I'm going to pass this New River Gorge National River bill today, you have a problem with that?" The district engineer must have agreed that there would be no impact to the Bluestone Dam, which was the upstream boundary of the proposed park unit, because the bill passed later that day.

Burton had many more legislative successes over the next four years, including another Omnibus National Parks and Recreation Act in 1980 that included numerous additional boundary expansions, including at Channel Islands and Biscayne, which were both redesignated as national parks. He secured wilderness designations in additional parks, including one I worked on extensively at New York's Fire Island National Seashore, and established new

New River Gorge National Park and Preserve rafting. Credit: Destry Jarvis.

historical units for African American sites in Boston, a Hansen's disease colony at Kalaupapa, Hawaii, Martin Luther King Jr.'s birthplace and family home in Atlanta, and the thousand-mile Ice Age National Scenic Trail of Wisconsin.

Perhaps no single member of Congress has had a greater impact on the composition of the national park system than Phil Burton. Over his six years as subcommittee chair, Representative Burton added more national park and wilderness acreage than all of the presidents and congresses before him combined. When all was said and done, he oversaw the creation of thirty new national park units, eight new national trails, eight wild and scenic river designations, and hundreds of smaller adjustments to parks and NPS authorities.

With President Ronald Reagan coming into office in 1981, Burton decided to leave the chair of the park subcommittee and instead chair a labor subcommittee, but he stayed very involved with parks through the fine subcommittee staff. He also hired NPCA's Joan Moody in December 1980 to be his press secretary and a legislative assistant in his personal office. As an editor of NPCA's *National Parks Magazine*, Moody had written extensively about parks in support of our program efforts and carried her deep knowledge of the parks and the threats to them into her new position.

After President Reagan took office in 1981 and James Watt became his highly controversial Secretary of the Interior, Burton took on a more aggressively defensive role, often excoriating Watt and other administration officials for their many adverse actions toward the NPS and other DOI agencies. At a March 1981 meeting between Secretary Watt and the NPS concessioners—the companies who operate the hotels and restaurants in the parks—Watt told them "If you have a problem with [the] NPS, come to me and we will change the policy or the person, whichever is easier." Watt also actively opposed new units of the national park system, publicly stating that the "NPS should take care of what it has before it reaches out for more." Watt also announced the "Park Restoration and Improvement Program" as an initiative to seek more funds for park facility maintenance, stating that there would be no new money for natural and cultural resources management in the parks, only for visitor facilities.

At one House Interior Committee hearing where I testified as part of a conservation witness panel, Burton complained loudly to Secretary Watt over his prodevelopment policies for the public lands, stating that he thought "the only way to deal with the polluters and the exploiters is to terrorize the bastards." A version of this quote, which he had also used earlier in the NPCA magazine interview, is literally carved into Burton's statue at Fort Mason in Golden Gate NRA. A terrific likeness of Burton sculpted by a former NPS employee, Wendy Ross, the sculpture features the quote on a card emerging from the pocket of his suit jacket, minus the "bastards" part. It seemed to me at the time that this was just what he had done to the timber companies clear-cutting the coast redwoods in his Redwood National Park Expansion Act of 1978; now Burton was turning his sights on the Secretary of the Interior. Over the course of Watt's tremendously controversial two-year term as DOI Secretary, I was a guest on several television news programs, including the *Today Show*, to rebut and counter many of his actions, or proposed actions, that would adversely affect the national parks.

After Mount St. Helens erupted in 1980, the drumbeat began to designate it as a unit of the national park system, like the other parks on the Pacific "Rim of Fire," including Lassen, Mount Rainier, Olympic, and Katmai. However, political appointees of the Reagan administration strenuously resisted transferring this US Forest Service (USFS) property to the NPS. Instead, legislation was enacted in 1982 designating Mount St. Helens National Volcanic Monument, to be managed by the Forest Service. For several years thereafter, the USFS built roads, trails, and several visitor centers to accommodate the public, in an effort to out-do the NPS. However, as is also typical of the USFS, its interest in managing sites for visitor education and interpretation waned, and operation of the monument's visitor centers and related programs were shuttered or transferred to Washington's State Parks in 2000.

Over the twelve years of the Reagan and George H. W. Bush presidencies, 1981–92, there were relatively few additions made to the national park system. Those that were added were not added at the request of the administration and were generally small historic sites or new units with peculiar or limiting authorities, especially ones that minimized land acquisitions. Included among new NPS units were the Natchez Trace National Scenic Trail (1983), which lies almost entirely within the already existing federal corridor of the Natchez Trace Parkway; Great Basin National Park in Nevada (1986), which transferred BLM and USFS land to the NPS along with the existing NPS-managed Lehman Cave National Monument; El Malpais National Monument in New Mexico (1987), again with BLM lands transferred to the NPS; Timucuan Ecological and Historic Preserve in Florida (1988), with heavy reliance on easements to protect lands and a management partnership with the City of Jacksonville; National Park of American Samoa (1988), with no federal land acquisition but managed through an agreement with the Samoan people; City of Rocks National Reserve in Idaho (1988) with a mix of BLM and state lands, comanaged by the NPS and the state; the Niobrara National Scenic River in Nebraska (1991), with only 3 percent of the land federally owned; Marsh-Billings-Rockefeller National Historic Park in Vermont (1992), all donated lands; and Little River Canyon National Preserve in Alabama (1992), also all donated lands, from a power company. Little River Canyon was one I worked on extensively, both with the landowner and his DC lawyer, Roy Jones, a former House Interior Committee professional staff member, and the local advocacy organization led by Pete Conroy. I represented the NPS leadership at the dedication ceremony for this new unit of the system.

While most of my work for the NPCA during the Reagan administration was focused in Washington, DC, whether on Capitol Hill or at the DOI, one important controversy took me to the field many times over a three-year period. As noted earlier, Acadia National Park in Maine first came into the system in 1916 as a national monument proclamation from President Woodrow Wilson, based on lands donated to the federal government for that purpose from a local nonprofit, Hancock County Trustees for Reservations. Between 1916 and 1982, Acadia had to change its boundary (a laborious process) fourteen times as additional lands were donated for the park. Acadia's authorizing legislation was unique, in that the NPS could not acquire lands except by donation, but it could, and did, accept donations anywhere across three counties, and not just on the islands. The end result was a patchwork of park lands interspersed with private lands that did not provide public access, making park management and visitor use extremely difficult in this heavily visited park. Local government officials, local and national conservation or-

Isle au Haut, Squeaker Cove, Acadia National Park. Credit: Destry Jarvis.

ganizations, the Maine congressional delegation, and the NPS all wanted a solution to this one-of-a-kind dilemma.

For the first, and I believe only, time in NPS history, the service recruited a professional mediator, provided by the Ford Foundation at no cost to the taxpayer, to negotiate a permanent boundary deal that would take passing a new law to achieve. It was decided that the deal would be worked out over a two-year period, principally by three individuals, one from the local communities, one from the NPS, and one from a nongovernmental conservation organization. I was chosen as the conservation representative, coming from NPCA. We essentially met on site every other month for much of the next two years, and we achieved agreement. The basic deal was that the NPS would get a fixed boundary set by law, within which it could purchase lands for the park, but would give up its authority to accept donations outside that boundary. As a result, the entire Maine bipartisan delegation in Congress, led by Senate Majority Leader George Mitchell (D-ME), introduced the required legislation. The three of us who had been the principals in the boundary negotiation all testified in favor of the bill. This was one park bill that the political appointees of the administration stayed out of, and it sailed through easily to become law in September 1986.

All of the new units the national park system added over the twelve years of the Reagan/Bush administrations were proposed by members of Congress. Under Reagan, NPS Director Bill Mott was allowed to testify in favor of several of these, but, more often than not, he stayed away from the Hill rather than appear to oppose the new parks. In contrast, during the George H. W. Bush years of 1989–92, NPS Director James Ridenour, previously Indiana State Parks director and a political selection by Vice President Dan Quayle from his home state, spent a great deal of his time railing against the "thinning of the blood" of the NPS caused by the addition of so many new sites for NPS to manage, without a concomitant increase in funding and staffing of the NPS.

In 1993, I went to work in the Clinton administration as a political appointee at the behest of incoming Assistant Secretary for Fish, Wildlife and Parks George Frampton, whom I had known and liked in his previous role as CEO of the Wilderness Society. Also, pushing for my selection into the position as NPS Assistant Director for External Affairs (covering Legislative and Congressional Affairs, Public Affairs, Partnerships, and Tourism) was Roger Kennedy, appointed as NPS director from his previous job as director of the American History Museum of the Smithsonian. In the second Clinton term, I moved up to the office of the assistant secretary for Fish, Wildlife, and Parks as senior advisor. For all eight years, my focus was on NPS policy and legislation.

DOI Secretary Bruce Babbitt came to Interior in 1993 with the priority of reforming the Bureau of Land Management and giving it a conservation mission to balance with its mining, grazing, and other multiple uses. Secretary Babbitt also wanted to push the BLM further by repeal of the 1872 Mining Act, by which private companies can file mining claims on public lands and gain ownership title at very little cost. And Babbitt wanted to reform grazing on public lands to at least charge ranchers the market rate for grazing leases on private lands. Babbitt was totally stymied by Congress on these two reforms and thus sought another course for BLM's conservation mission.

The first major legislative issue facing the Congress and new Clinton administration that brought BLM reform into focus was the California Desert Protection Act, which was being pushed hard by most of the California delegation, especially Senator Dianne Feinstein. In 1976 Congress had designated the 25-million-acre California Desert National Conservation Area as a BLM special management area entirely as a multiple-use management zone. National conservation organizations, including the NPCA, Sierra Club, the Wilderness Society, and others, wanted much greater conservation protection for key sites in the desert than BLM management would allow.

In the 1980s, before I left the NPCA, I had hired a park ranger away from the NPS, Brien Culhane, to add to NPCA's conservation staff. He proved to be a smart and tenacious policy advocate. Soon after I assumed my NPS position in 1993, I hired Brien into the NPS Legislative Affairs office, and he quickly took the lead role on my staff for this California legislation. The California

Desert Protection Act became law in October 1994 and established Mojave National Preserve as a unit of the NPS with the transfer of 1.5 million acres of BLM lands. In addition, Death Valley and Joshua Tree National Monuments, longtime NPS units, were redesignated as national parks and expanded with the transfer of 1.3 million acres from BLM to Death Valley National Park and 230,000 acres to Joshua Tree National Park.

With that NPS vs. BLM battle behind him, Babbitt determined that he could attempt to build a real conservation mission and ethic into the BLM by having President Clinton proclaim national monuments on important BLM public lands, and then retain them under BLM, rather than NPS, management. Under the Antiquities Act, the president alone may proclaim a national monument from among the public lands or lands donated for that purpose. Early in his tenure, President Clinton officially asked DOI Secretary Babbitt to prepare national monument proclamation proposals for his consideration; but clearly the idea of designating most of these for continued management by the BLM was Babbitt's.

However, each agency that has been given national monuments to manage does so under the basic laws that control that agency. For the NPS, that is the 1916 NPS Organic Act, with its clear management priority set on preservation. For the BLM or USFS, monuments under their management are still controlled by each agency's underlying statutes, which for both agencies is a multiple-use sustained yield mandate. Thus, the USFS still cuts timber in Giant Sequoia National Monument, for example, and the BLM still leases its national monument lands for livestock grazing and mining.

Over the eight years of the Clinton administration, the president proclaimed nineteen national monuments and expanded three others that already existed, covering some 3.5 million acres, most of which were already managed by BLM for multiple uses. Among these monuments, two are intended to be managed jointly by the NPS and BLM: Grand Canyon Parashant, in Arizona, was designated on lands already within Lake Mead National Recreation Area, and Craters of the Moon in Idaho. Each was an existing NPS unit that was expanded by the proclamation, alongside newly designated BLM monument lands.

Both during the Clinton years and for some years thereafter, I worked to get recognition and protection for all of the World War II Japanese American internment camps, and I drafted the Clinton Proclamation for the Minidoka National Monument, which he designated in 2001. I discovered an NPS internal historical research document entitled *Confinement and Ethnicity*, by Dr. Jeff Burton, on the internment camps, and had it published and widely disseminated. This is a beautifully written and fully documented account of this mistaken and shameful period of our history: one not to be forgotten now that

Manzanar National Historic Site, World War II Japanese American Internment Camp.
Credit: Ansel Adams, photographer; Library of Congress, Prints and Photographs Division.

four of the camps, Manzanar, Minidoka, Tule Lake, and Honouliuli, are units of the NPS, and all the others are designated as national historic landmarks.

Late in President Clinton's first term, I worked closely with congressional committee staffers and Congress to pass the significant Omnibus Parks and Public Lands Management Act of 1996. This act added six new units to the system, including several that NPS Director Roger Kennedy strongly pushed—expanding a trend to bring more representation of minority achievements into the system. Added then were New Orleans Jazz and Cane River Creole sites in Louisiana; Nicodemus, a post–Civil War African American historic townsite in Kansas; and significant civil rights sites, including the Selma to Montgomery National Historic Trail and the Tuskegee Airmen National Historic Site in Alabama and the Little Rock Central High School in Arkansas. These diverse sites followed closely on three created in the latter part of the George H. W. Bush administration, including the Mary McLeod Bethune Council House in Washington, DC, the Manzanar Japanese American Internment Camp in California, and the Brown v. Board of Education site in Kansas, commemorating the landmark and precedent-setting Supreme Court civil rights case won by NAACP attorney and later Supreme Court Justice Thurgood Marshall.

Four significant, trend-setting changes for NPS were initiated in the 1996

Omnibus Act: First, the late Representative Phil Burton had set the stage for addition to the NPS of the historic Presidio Army Fort in San Francisco, if it were to be closed. Congress added management of the Presidio coastline and beaches to Golden Gate NRA, while authorizing the Presidio Trust, a quasi-governmental corporation, to manage adaptive reuse by leasing the historic buildings of the fort.

Second, the act authorized the Shenandoah Valley Battlefields National Historic District (NHD), covering eight counties in the Valley of Virginia that featured significantly in the Civil War. One site within this district, Cedar Creek Battlefield, was made an NPS unit, but any future preservation and management of the dozens of other historic sites within these largely rural farming counties was left to a newly authorized "management entity," the Shenandoah Valley Battlefields Foundation (SVBF). SVBF was given a line item for land acquisition funding, authorization for an appropriation for operation, management, and small grants to support other nonprofit initiatives in the Valley NHD. This provision of the act represented a new approach to large landscape conservation, relying on a congressionally established nonprofit foundation partnership to achieve most of the goals of the statute for this NHD. This model has been a success, in part, due to innovative directors of the foundation, including its current CEO Keven Walker, who was hired away from the NPS.

Third, the act separately authorized nine new national heritage areas (NHAs). The first NHA, the Illinois and Michigan Canal, had been authorized in 1984 during the Reagan Administration, which often referred to it as a "new model national park"—that is, one with no federal lands or direct NPS management. NHAs are designated by Congress as places where natural, cultural, and historic resources combine to form a cohesive, nationally important landscape. NHAs are lived-in landscapes, where no new federal land acquisition is authorized, and where day-to-day coordination is through a local nonprofit entity. By 2020, Congress had designated sixty-nine NHAs in thirty-four states. Several of them have one or more small NPS unit(s) within a much larger landscape boundary, but all still rely on state, local, and private support, rather than being fully federally funded. They are not considered part of the national park system and as yet have no overarching "organic act" to define their mission and management goals.

Fourth, the act authorized the Tallgrass Prairie National Preserve in Kansas, with only a historic ranch house to be owned by the NPS within a 10,000-acre property owned by a nonprofit organization. Since at least the 1960s, the NPS had argued that the last major physiographic province of the United States that had no representation in the national park system was that of the tallgrass prairie. In the late 1980s, it had looked like a large ranch for sale in Oklahoma would win the political battle over securing a tallgrass prairie unit for the system.

The Oklahoma state director of the Nature Conservancy (TNC), John Flicker, and I arranged a visit to several national parks for a number of key local leaders from this area of Oklahoma, including the owner of the ranch that was for sale, the chair of the local Osage Indian Tribe, the elected county leader, and several local conservation activists. We visited Badlands National Park, Grand Teton National Park, and Chickasaw National Recreation Area on a four-day tour. Upon our return, all parties agreed that the Oklahoma tallgrass prairie would be an ideal NPS unit and to push for the needed legislation. Unfortunately, the Oklahoma congressional delegation (all Republican) would not agree to support the bill. TNC went ahead and purchased the 15,000-acre ranch and continues to manage it today as a private tallgrass prairie preserve.

After I left NPCA in 1988, the staff lead there moved to Laura Loomis, who tenaciously pursued political support for a NPS unit in the tallgrass prairie until it was successful. The NPCA negotiated with a wealthy philanthropist and cattle rancher to donate the funds to purchase the Spring Hill Ranch in Kansas, with the proviso that he would retain the grazing rights on the property.

In 1997, as NPS assistant director, I was the NPS representative sent to Kansas for the dedication ceremony of the preserve after the authorization was passed with the leadership of Senator Nancy Kassebaum (D-KS). It was a beautiful clear fall day as Kansas Governor Bill Graves, Ed Bass, the philanthropist/rancher, and I walked together to a hilltop for the ceremony. As we moved among the gently swaying grasses, Mr. Bass quietly remarked to us that Congress should not have named this area the Tallgrass Prairie National Preserve, since his cows would eat all the tall grasses. Instead they should have just named it Big Vista National Preserve. I was stunned by the callous comment totally counter to the vision for the place that its many supporters intended, and I made sure to include in my remarks at the ceremony that the tall grass ought to be allowed to flourish and that bison should replace cattle. Fortunately, in 2005 the Nature Conservancy bought out the grazing rights. Today, the NPS and TNC enjoy a management partnership there, part of a cooperative public-private trend that is continuing today in a growing number of newer NPS units.

While President Clinton and DOI Secretary Babbitt achieved a strong conservation record over their eight years, it was not at all a smooth and quiet time. In particular, shortly after the midterm election in fall of 1994, Newt Gingrich (R-GA) became Speaker of the House in a significant flip of the majority. The so-called Republican Revolution flipped fifty-four House seats, giving that party the House majority for the first time since 1952. Very early in the

104th Congress, Speaker Gingrich issued his major federal reform proposals, which he called a "Contract with America," including massive budget cuts and wholesale federal government restructuring. For the NPS, this meant big trouble. I called it a Contract *on* America, and I and my staff at NPS worked against its proposed impacts on the NPS for the next several years.

Having seen this major realignment of Congress looming, early in 1994 I arranged for Director Kennedy to send a memo to all park superintendents asking them to prepare and submit to us a "Statement of Significance"— a clear and short statement on why each unit of the system was nationally significant. Each of these 368 statements of significance was written by the unit superintendent and his/her career professional staff in order to best reflect the knowledge and scholarship, as well as congressional intent, that represent the purposes of that unit.

Late in 1994, as we approached the start of the 104th Congress, I began to organize an eager group of twenty NPS career staff from across the field units and programs of the service to come to Washington for a period of weeks as a "Bridge Team." They all arrived in early 1995 as soon as the new Congress was seated. These top NPS staff were not in DC to lobby Congress on any pending legislation or for increased funding. Their task was simply to orient all these newly elected members and their newly hired staff to the NPS mission and to which units and programs of the service were serving their congressional districts.

I had the NPS Interpretive Design Center prepare a beautiful but simple four-color booklet entitled *Caring for the American Legacy: The National Park Service.*[8] This booklet was designed with pocket sleeves into which we could insert individualized sheets tailored to each member's district. Every office of Congress received a visit from a member of the Bridge Team and a packet specific to that district, including the Statement of Significance for each park there. This effort was helpful but did not deter the new House leadership from seeking to dismantle major elements of the national park system as they moved quickly to set up a "park closing commission," aimed at removing units of the system that did not receive high visitation. Likewise they sought major cuts in the NPS budget, which would then enable them to assert that the NPS could not take care of the system. They ultimately failed.

For some years, while in the minority, Utah Representative Jim Hansen (R) and Colorado Representative Joel Hefley (R) sponsored legislation to establish a commission to study deauthorization of units of the national park system that were judged *by them* to be either not nationally significant or as having too few visitors to deserve the designation. While the Democrats were in the majority, none of these bills had ever received a hearing, much less been enacted, due to strong opposition. Late in 1994, Representative Hansen was cited in newspaper articles discussing his intent, if he became the next chair of the House Parks Subcommittee, to remove some units from the system. He said, "Next Tuesday [referring to the November election], if we get more business oriented people, I think we'll see some dramatic things occur. . . . If we take over, we're going to do a Parks Closing Commission."[9]

At the beginning of the 104th Congress, Jim Hansen became chair of the House Natural Resources' Subcommittee on Parks and Public Lands. Chair Hansen and Representative Hefley, on January 5, 1995, introduced HR 260, the "National Park System Reform Act," which directed the DOI Secretary to "develop a report which contains a list of areas within the national park system where National Park Service management should be modified or terminated." The bill further specified that if the Secretary failed to do the report to Congress (likely, given that the Clinton administration was adamantly opposed to the bill), then a National Park System Review Commission would be established, consisting of seven members, three appointed by the Secretary and four appointed by Congress, with the stated goal of producing a list of existing park units that should be modified, transferred to another agency, or abolished.

NPS Director Kennedy testified at a subcommittee hearing in late February, expressing NPS and administration opposition to HR 260, but noting "We support instead a forward looking measure which would give the NPS the opportunity to review its existing set of criteria for *future* sites under consideration for inclusion in the system."[10] Along this line, former subcommittee chair and then ranking Democrat Bruce Vento (MN) introduced an amendment that would require the NPS to do proposed new area studies only after each was specifically authorized by Congress. This new area study process was separately enacted several years later by Congress as section 301 of the 1998 Omnibus National Park Management Act (Public Law 105-391).

Facing strong opposition to HR 260 from the administration and many nongovernmental organizations, when the full Natural Resources Committee met in May to mark up the bill, instead a substitute bill was adopted, which precluded review of the fifty-four statutorily named "National Parks" but allowing the review to include the remaining 314 units of the national park system. NPS Director Kennedy sent the committee a letter on May 9, 1995, that I drafted, continuing opposition to HR 260 and stating that "the Administration continues to oppose the privatization or divestiture of any national park unit." Nevertheless, the full committee reported the bill, as amended, favorably on June 5.

From the outset, advocates for the park closing commission had based their support for eliminating parks on the inadequate budget given to the NPS to fully manage all the units of the system in a high-quality manner.

They argued that the number of recreation visits should be a key factor in determining what qualified for inclusion in the system—the fewer visitors, the less qualified a site ought to be. They also argued that the big, green, natural-area parks, mostly in the West, were the "real" or "good" national parks, and the small historic sites or urban recreation areas did not qualify to be in the national park system. Ironically, and tellingly, no member of Congress ever cited an NPS unit in his/her own district that ought to be closed—those are just too popular: the bad ones were apparently in another member's district.

Furthermore, the inadequate budget argument did not hold up to close investigation, primarily because it coincided with major federal budget cuts being proposed by Speaker Gingrich, including a 20 percent cut to the NPS budget, increasing to a 40 percent cut by 2000. In order to achieve this level of budget cuts, the NPS would have to close some two hundred units of the system based on their annual appropriations for operations. At that time, the total NPS appropriation represented some 0.066 percent of the federal budget. Closing all twenty-seven units that had been added to the national park system since 1981 would only reduce the NPS budget by 3 percent, and that did not account for the fact that each unit cut out would have to be transferred to some other entity—federal, state, local, or private—that would still have a similar cost to bear.

On September 19, 1995, Committee Chair Don Young and Representative Hansen asked for a vote of the full House on HR 260 under "Suspension of the Rules," a rule that allows no amendments on the floor but requires a two-thirds majority to pass. The bill failed on a vote of 180 Yea–231 Nay. A few days later these members tried to add HR 260 to the Budget Reconciliation Act, which also failed. Committee leadership made only feeble attempts to resurrect this idea again, and the Omnibus Park and Public Lands Act of 1996 passed with wide support the next year, without such a park closing commission even considered.

During George W. Bush's first term, his Secretary of the Interior, Gale Norton, and her team of appointees from the oil and gas, mining, and timber industries attempted to systematically reverse many of the conservation actions of the Clinton administration. In January 2002, she announced a review of twenty-two national monuments to consider reducing their boundaries in order to permit mining and other developments, and she announced plans for a major reduction in the NPS budget. By July Norton announced plans for oil and gas leasing directly off the Florida coast, threatening Everglades National Park, as well as to open the coastal plain of the Arctic National Wildlife Refuge for oil and gas drilling. In November, she ordered the Everglades Restoration Office in Florida closed.[11] All of these efforts on her part ultimately failed, facing strong opposition from members of Congress, numerous conservation organizations, and myself in my new capacity as a nonprofit consultant, working on these issues for the NPCA, the Wilderness Society, and others.

Secretary Norton's main contribution to the policies of the Department of the Interior, other than consistently denigrating the NPS's preservation mission, was to espouse the idea of "cooperative conservation." This phrase sounds like a fine idea for coordinating management over large landscapes, but her definition of it was simply to give local governments and federal ranching and mining lease holders more control over federal land management at the expense of natural and cultural sites. In the end, this effort also failed, and Secretary Norton was replaced in George W. Bush's second term by a far more park-friendly Secretary, Dirk Kempthorne.

During President George W. Bush's eight years in office, there were only four units added to the national park system, all small historic sites—Flight 93, Carter Woodson House, Cedar Creek Battlefield, and World War II Valor in the Pacific National Monument.

JONATHAN

The National Park Service often assembles teams from the parks to work on the details of potential new national park designations, generally at the behest of Congress or the President. While serving as the superintendent of Craters of the Moon in the early 1990s, I had developed a good working relationship with the Bureau of Land Management, which manages public lands bordering the park. This must have been the reason I was tapped to serve on a planning team to recommend the details and boundaries for a new unit of the NPS in California's Mojave Desert. As a member of a small team led by Mammoth Cave National Park Superintendent Dave Mihalic, we spent days traveling the California desert and evenings over beers and maps. Although we had a great BLM representative on the team, it was clear that the BLM leadership was opposed to any new designation that would take "their" lands and give them to the NPS. However, they did recognize this effort was being driven by Senator Dianne Feinstein (D-CA) and was not to be stopped, so they had better make the best of it.

I prepared a report for the team on the issue of hunting and recommended against allowing it in the new park. The politics were too strong from the BLM and the California Fish and Game, so hunting was going to be permitted; therefore, our team recommended that the new unit carry the title "pre-

serve" rather than "park." As we finalized our recommendations for Director Roger Kennedy over a few beers in Las Vegas, team leader Mihalic told
us that he had once sold kitchenware door-to-door to pay for college. He said
he developed a trick, where he would hold the contract for buying his set of
kitchenwares along with a pen. As he explained the contract to potential buy-
ers he would let the pen slide down the contract, so that they caught it. He
knew that once they had the pen in hand, they would sign. We dared him to
do this with Director Kennedy, who needed to sign the one-page agreement
we had crafted between the NPS and the BLM detailing our team's recommendations of land transfers. Never one to back down on a dare, Dave held
the agreement in front of Director Kennedy as he explained it, then skill-
fully let the pen slide, and Kennedy caught it. Dave kept talking and laid the
agreement onto the conference table and Kennedy signed it and then looked
up and said, "What did I just sign?" In October 1994, the 1.5-million-acre
Mojave National Preserve was established based on our recommendations.

During the administration of George W. Bush, I was called to Washington
to brief the NPS leadership on the proposed expansion and redesignation of
Fort Clatsop National Memorial to celebrate the 200th anniversary of Meri-
wether Lewis and William Clark and their Corps of Discovery. Local support
as well as that of the Washington and Oregon congressional delegations
seemed to make this legislative action easy, but the Department of the Interior, under Secretary Gale Norton, was opposed. After briefing Randy Jones,
the career deputy director of the NPS, he said I would have to go "upstairs"
alone to brief the Secretary, as he said "they" "hated" him and he could not
help, but first I had to brief David Bernhardt, the counselor to the Secretary.
(Bernhardt would become Secretary of the Interior under President Donald
Trump.) I sat down in front of Bernhardt and started my briefing with a discussion of the politics. He cut me off and said that politics was his job, not
mine, and then he posed the question of why he should support a new park
from me when he had told every member of Congress they could not have a
new park. As I started to explain the historical importance of the expedition,
he cut me off again and said he did not want to hear anything about Lewis
and Clark. Well aware that this was not going well, but that the legislation
had momentum, I just said, "The parade for this park has already started.
You can stand on the sidewalk and watch it go by, or you can get in front and
claim credit, your choice." Bernhardt said "good point" and sent me to brief
the Secretary in the adjacent room. I presented the proposed legislation, local input, history, and maps of the boundary changes to Secretary Norton.
She sat quietly and listened but never asked a single question and I was dismissed. A few weeks later, the department issued a press release supporting
the designation of Lewis and Clark National Historical Park.

When I was sworn in as the director of the National Park Service on October 2, 2009, I had several goals in mind. Drawing from the vision, goals, and accomplishments of my predecessors, I refined my focus into four broad areas: education, relevancy, stewardship, and workforce. Sometimes referred to by employees as the "four pillars of Jarvis," these categories guided my eight years in office.

I chose education as a focus because, from my early days as a ranger, I knew that the public loved to learn about the parks, their histories and back-stories. I also recognized that the NPS had never been viewed as an educational institution, yet millions of visitors learned about complex issues such as fire, glaciers, wildlife, and the Civil War from well-trained rangers leading walks and evening campfire programs.

I chose workforce because I knew that the NPS had a terrible safety record, and we needed to change the culture to prevent employee injuries and death on the job. I also knew that employee morale was low, due in part to the treatment of career employees by the previous administration of George W. Bush.

I chose stewardship because, at the core of the NPS mission, preservation of the resource had taken a backseat to visitor services and, with the looming challenge of a warming planet, the parks were highly vulnerable to climate changes.

And I chose the term "relevancy" because the diversity of the nation was not reflected in the NPS workforce, park visitation, or the portfolio of parks. If the national parks are not relevant to each succeeding generation, then they will slowly decline through disinterest and neglect. I also noted that visitation had been flat for nearly a decade, holding steady at around 280 million. In my first congressional testimony to justify the NPS budget, I was asked repeatedly why we needed an increase in the budget when visitation had not increased. I knew this to be a false correlation, as the management of the parks is much more complex than just managing visitors. But I also knew that the ability to fund the national parks was directly related to their political support, which, at least in part, stemmed from a positive park experience.

Drawing on the history of the NPS for inspiration, I looked back to the post–World War II period. Then the national parks had been neglected to the point of embarrassment, leading writer Bernard DeVoto to pen "Let's Close the National Parks" in *Harper's Magazine* in 1953. The Mission 66 effort, launched by NPS Director Conrad Wirth and carried forward by Director George Hartzog, was more than an infrastructure program. It was also a marketing effort, targeting the returning World War II veterans and their families to see the nation. With partners in the automobile industry and the catchy melody "See the USA" by Dinah Shore, the public was invited back to their parks. And they came in droves. From that reconnection to our parks

and public lands, public support led to the passage of the bedrock conservation laws. During this period of progress, Congress passed the Wilderness
Act (1964), Land and Water Conservation Act (1964), National Historic Preservation Act (1966), Endangered Species Preservation Act (1966; amended
as the Endangered Species Act in 1969), Wild and Scenic Rivers Act (1968),
National Environmental Policy Act (1970), and the Clean Air Act (1970). I
felt I could create a new marketing effort that would instill a similar sense of
stewardship and patriotic pride as had been done in the 1960s and 1970s.
This thought, combined with the need to reach a new constituency, moved
me to partner with the National Park Foundation and launch the "Find Your
Park" campaign, specifically designed to connect with the more diverse mil-
lennial generation. The goal was to inspire them to visit a park—and become
supporters and advocates. I also saw this as a way to build a deeper and more
resilient body of public support for the parks that would come to their defense
should another hostile administration come into power. I also knew there
would be a lag in the response by Congress, but that renewed public support
would translate into additional funding in future years.[12]

The 1916 Organic Act establishing the NPS stated the parks would be
"promoted and regulated . . . by such measures and by such means" that
they would be unimpaired for the enjoyment of future generations. I interpreted this to mean that the parks we manage and the stories we tell and the
employees we hire must reflect the diversity of the nation. However, social
science on public attitudes to conservation and demographic data from visitation demonstrated that national park visitation and support tended to be
from the older, white, middle-class part of the population. I had the opportunity to change that, or at least to make a significant contribution. We could
emphasize stories that built relevancy within the existing system, such as
the African American Buffalo Soldiers and their stewardship of parks like
Yosemite at the turn of the century. With the upcoming sesquicentennial of
the Civil War, we could interpret how slavery was the cause of the Civil War
with revised ranger programs at the battlefields of Manassas, Gettysburg, and
Antietam. But from my perspective, there were also parts missing from the
system: What better way to build relevancy than to engage in finding those
places and adding them to the stewardship of the National Park Service?

While moving new park units through the legislative process is always
a possibility, it tends to be slow and subject to all the standard objections
from conservatives: take care of what you have, can't afford new parks, and
so on. Plus, stand-alone park bills rarely have the political weight to move
independently through Congress and often have to wait for a "carrier" bill
that has other benefits for bipartisan support. With the election of Barack
Obama in 2008, the Democrats controlled both the House and the Senate,

but that was short-lived, as the Republicans took control of the House in the midterms of 2010. Under Republican leadership, the committees overseeing the NPS demonstrated hostility to the mission of the service, and any legislative proposal for expansion died quickly. I was regularly dragged before these committees and grilled over basic conservation issues.

The alternative to legislation is to follow the route of the 1906 Antiquities Act, using the exclusive power of the president to designate new national monuments. Nearly every president since Teddy Roosevelt has used the Antiquities Act during his tenure, with a few exceptions: Richard Nixon, Gerald Ford, Ronald Reagan, and George H. W. Bush. Because I was working for the first African American president in our nation's history, I knew that President Obama would be interested in supporting new additions that would build relevancy. I also knew that Secretary of the Interior Ken Salazar was passionate about diversity, as he had suffered discrimination as a young Hispanic growing up in a rural, Mormon-dominated area of Colorado. I canvassed a range of outside individuals, including Destry, and developed a working list of potential new areas that could be added. I carried that list with me at all times, working the angles every day to capture as many of them as possible.

The first success was one that I had started before coming to Washington. Port Chicago Naval Magazine is the site where on July 17, 1944, two navy munitions ships exploded, killing 320 men in the largest World War II homefront disaster. Most of the men were African American, and when those who survived refused to go back to work without better safety systems, they were court-martialed and imprisoned. Observing the trial was the young attorney representing the National Association for the Advancement of Colored People, Thurgood Marshall. The injustice of this tragedy initiated the desegregation of the military. Working with Congressman George Miller, we were able to get a bill through Congress that designated the site as a national memorial under the stewardship of the National Park Service on October 28, 2009. Secretary Salazar commented to me that I had only been officially on the job for three weeks and already had added a new park to the system.

I also had a long list of natural areas I would have liked to add to the NPS; these historically had been carved out of one of the other federal land management agencies, the US Forest Service or the Bureau of Land Management. The Forest Service had long resisted these attempts, viewing such legislative actions as pure thievery and an insult to their own (multiple-use) stewardship. The beltway animosity between the USFS and the NPS goes back to differences in conservation ideology between conservationist John Muir and the first chief of the US Forest Service, Gifford Pinchot. The net result is that the NPS and the USFS are in different departments of the executive branch and remain competitive. Discussions between our agencies in the early days of the Obama administration indicated that there would not only be no support but active resistance for creating national park units out of US Forest Service lands.

New park units had also been carved from the vast holdings of the Bureau of Land Management, also part of the Department of the Interior, but the culture of the department had shifted during the tenure of Secretary Bruce Babbitt. Secretary Babbitt wanted to give the BLM more of a conservation ethic by charging it with the stewardship of new national monuments under the umbrella of a National Landscape Conservation system. Since his tenure, most of the new conservation designations under the Antiquities Act have remained under BLM (and USFS or USFWS) management. The success of that approach remains mixed, as the BLM and the USFS are multiple-use agencies, charged with extraction as well as stewardship, and are often forced to lean to extraction as the priority. So I picked my battles with my fellow agencies carefully. There were also political appointees from the Babbitt years who came back to Interior with their bias against the NPS, viewing the agency as too independent, needing tight control, and "too big for its britches." In addition, as detailed by Destry above, the Office of Management and Budget was the official buzz killer of any suggestion of expansion. To achieve additional parks and to get them funded, I would have to take on both OMB and internal opposition at the Department of the Interior directly—or go around them straight to the president.

Rather than blunting my sword on land transfers from other federal agencies, I chose to focus on historical sites that represent the contributions and struggles of women and people whose civil rights have been won only through long and persistent fights. Historically, presidents use the Antiquities Act to create new national monuments late in their terms, often as they go out the door. In 2009, Secretary Salazar launched the America Great Outdoors and Treasured Landscapes program and asked for my priorities. I outlined new wilderness designations as well as at least ten new historical and natural parks, along with boundary adjustments to existing parks. Taking a cue from missteps during the Clinton administration and a perception that Antiquities Acts designations were done without public input, Salazar committed that, before any such action by the president, there would be public involvement and at least one public meeting. This became the standard, and I not only attended but also sat on the stage to take the public comments on all of the NPS monuments created during the Obama administration. I actually like public meetings, as they are an opportunity to really hear from a broad range of the public.

I also knew I would need help, so first I contacted a former colleague, Molly Ross, who had served as a solicitor in the Department of the Interior

National Park System Advisory Board at Mesa Verde National Park. Credit: Jonathan Jarvis.

for decades and had recently retired. I asked Molly to return as a "retired an-
nuitant" and work exclusively for me on Antiquities Act monuments. Second, to ensure that I was focused on the right set of stories, I charged the
National Park System Advisory Board with helping me uncover the contributions of women and minorities in America. The National Park System Advisory Board was established by Congress in 1935 to be comprised of citizens
with special expertise to advise the director on the designation of National
Historic and Natural Landmarks. It also serves as a sounding board for the
director on major topics and, by law, conducts its business before the public.
Early in my tenure in 2010, I had reconstituted the National Park System
Advisory Board, hand-selecting twelve new members, most of whom had already served on the Second Century Commission.[13] (I had served as the official NPS liaison to the commission and had come to know and respect the
commissioners. They had already spent two years getting to know the inner
workings of the NPS, so it was logical to pick the best of the best to serve on
the advisory board.) While the advisory board has a specific legislated responsibility to nominate national natural and historic landmarks to the Secretary,
I expanded their scope of work to include science, education, workforce, leadership, diversity, and philanthropy.

With the strong guidance of NPS Associate Director for Cultural Resources and Science Dr. Stephanie Toothman as well as advisory board members Tony Knowles, Milton Chen, Belinda Faustinos, Carolyn Finney, Stephen Pitti, and others, we began a series of discussions and scholarly reports[14] to recommend the most important sites for designation as national historical landmarks or other units of the national park system.

I also wanted to get a political winner on Obama's desk early in his tenure, leaving enough time to pursue others while he was still in office. Fort Monroe became that first effort.

The Fort Monroe story is rich and compelling. One summer day in 1619, a ship appeared off what was then known as Point Comfort, the location of an English fort overlooking the Chesapeake Bay. That ship would later come to be called the "African Mayflower" because it carried the first enslaved Africans to the colonies. Trading with food, the colonists took a few Africans as slaves. One was a pregnant woman, who soon gave birth to the first enslaved person of African descent born in what became the United States.

By the time of the Civil War, Point Comfort had become a Union stronghold known as Fort Monroe, the only Union fort south of the Mason-Dixon Line that never fell to the Confederacy. In the middle of the night, on May 27, 1861, three escaped slaves, Shepard Mallory, Frank Baker, and James Townsend,[15] appeared at the fort's gate, looking for sanctuary. General Benjamin Butler was in command, and when the Southern slave owners demanded the return of their property, Butler refused, in spite of the fact that at that time slavery was legal in the United States.

Butler, an attorney by training, reasoned that the slaves were Confederate property, used in the execution of the war by the enemy, and therefore could be declared contraband and confiscated by Union forces. He "confiscated" these three individuals and gave them protective custody, and, when word spread, his practice became a federal policy known as the Contraband Decision. President Abraham Lincoln traveled down to Fort Monroe and spent the evening with Butler, where we assume they traded their legal views of this decision, perhaps over a brandy or two.

Lincoln returned to Washington and penned the first draft of the Emancipation Proclamation. The three Fort Monroe fugitives were the first slaves freed in the Civil War and many more would follow as a result. Fort Monroe thus bookends the beginning and end of slavery in the United States.

The protected encampment of slaves outside the gates of Fort Monroe grew to become today's city of Hampton, Virginia, and the mayor, Molly Ward, contacted me (through Destry), suggesting that the NPS take a look at Fort Monroe, which was finally undergoing a base closure and disposal by the military. I drove down and met Mayor Ward on site and we toured the old military grounds. There was the original "star fort" and its characteristic moat, and one of the cells had been used to imprison Jefferson Davis, the president of the Confederacy. Much of the original buildings remained intact, including the quarters of Robert E. Lee, who as a US Army engineer lieutenant had overseen construction of the fort, and the offices of Benjamin Butler, where he hosted President Lincoln. The fort rested at the foot of the Chesapeake Bay, with a long line of open beaches that would have been seen by Captain John Smith as he explored the bay in the seventeenth century. It was perfect for Antiquities Act designation.

Although Virginia is conservative, and the local congressman at the time was a self-professed Tea Party member, Mayor Ward assured me there was lots of local support for the National Park Service taking at least part of Fort Monroe. My team began to research the base closure and discovered that when the fort had been established as a military site, the Commonwealth of Virginia had retained a reversionary clause for the underlying lands to return to the state, should the military leave. This meant it would take an act of the Virginia State Legislature and support from the governor to allow the NPS to take possession. We had our work cut out for us, including complex issues of the military transition, multiple buildings that would need to be rehabilitated, a dump site with contamination, a commercial marina, and so on. Over the next two years, Molly Ross, Mayor Ward, the solicitors in the department and the Council on Environmental Quality (CEQ), the Secretary, and the White House worked every angle. I met personally with Governor Robert McDonnell (who later went to prison for corruption), sparring over which one of us had deeper Virginia roots. After Governor McDonnell explained his family's arrival in Virginia was in the 1800s, I just said John Jarvis got off a ship at Jamestown in 1620. He said, "I can't touch that."

In July of 2011, we hosted the required public meeting in the officer's club at Fort Monroe. We estimated the crowd to be over a thousand. Secretary Salazar conducted the show. I have witnessed many politicians handle large crowds, and Salazar was one of the best. He was fair, funny, engaged, respectful, and always had command of the room. Near the end of the meeting, he asked everyone who supported the idea that President Obama would use the Antiquities Act to designate a unit of the national park system at Fort Monroe to raise their hand. Hands shot up, nearly unanimous, including the local Republican congressman.

On Tuesday, November 1, 2011, I walked over to the White House and was escorted into the Roosevelt Room, gathering with Secretary Salazar, Mayor Ward, and others. The door to the Oval Office opened and President Obama warmly greeted us. As we stood around his desk, he signed the presidential proclamation, under the authority of the Antiquities Act, establishing Fort Monroe National Monument as the 396th unit of the National Park Service.

Jonathan Jarvis with President Obama. Courtesy Barack Obama Presidential Library.

Before we left the office, he turned to me and said "You got any more of those?" I said, "Yes sir, I have a long list."

With that kind of support from the president and a memo from his office asking for recommendations—and in spite of constant resistance from the budget hawks at the Office of Management and Budget—our little team really kicked into gear to bring new areas of historical importance up for designation by the president. We had a system that started with a one-page brief that identified the importance of the site, the ownership, the politics, and the potential of acquisition. I would run that list by the Secretary, first Ken Salazar and later Sally Jewell. With a green light, the legal team would begin its analysis, and a ground team, usually from a nearby unit of the NPS and with local advocacy groups, would begin to plan the public meeting. We would also engage the congressional delegation and suggest they at least attempt moving a bill; but with Congress in a deadlock and the conservatives working hard to undermine the president, legislation was not moving. I also did not want to overwhelm the NPS with new responsibilities, so I worked with the NPS comptroller and budget staff and the House appropriations committee to ensure there would be at least a base of funding support. In addition, the National Park Foundation was on the rise and capable of facilitating some philanthropic support for the new units.

Our research found the most important sites across the nation that would broaden and deepen the historical narrative beyond just the contributions of white men. Along the way, I too learned a great deal about our nation's history of racism, discrimination, and the valiant people who resisted. I knew the NPS had a responsibility to tell these stories.

With humble beginnings as a migrant farm worker, César Chávez studied the peaceful activism of Dr. Martin Luther King Jr. and Mohandas Gandhi and began in the 1960s to organize the Hispanic and Filipino farm workers of California and Arizona. Legally barred from union representation, Chávez organized peaceful marches, protests, and boycotts that drew national attention. His own personal sacrifice by fasting drew the attention of national leaders including Robert Kennedy. In spite of massive political resistance from the farm owners, his efforts, along with those of other leaders in the movement, resulted in the organization of farm workers as the United Farm Workers Union in 1962. Their efforts improved the living and working conditions for millions of people who still today work in our fields.

Secretary Salazar and I conducted the public meeting for the potential designation of a monument to César Chávez. Chávez's family, including his widow and his son Paul, were cautious about trusting Chávez's legacy to the government, so they traveled as a group to Manzanar National Historical Site and learned how the NPS treated the racially motivated confinement of Japanese Americans during World War II. This experience convinced them that they could trust the NPS with César's story. On October 8, 2012, on the campaign trail for a second term, President Obama came to La Paz, California, and addressed an adoring crowd. To chants of "si se puede," the president spoke:

> More than anything, that's what I hope our children and grandchildren will take away from this place. Every time somebody's son or daughter comes and learns about the history of this movement, I want them to know that our journey is never hopeless, our work is never done. I want them to learn about a small man guided by enormous faith—in a righteous cause, a loving God, the dignity of every human being. I want them to remember that true courage is revealed when the night is darkest and the resistance is strongest and we somehow find it within ourselves to stand up for what we believe in.[16]

And with a stroke of his pen, President Obama designated Chávez's home and office in La Paz as César E. Chávez National Monument, part of the national park system.

During its struggle for independence, in a colonial courthouse at New Castle, Delaware, this nation set itself on a course unprecedented in the world. It was here that Delaware ratified the constitution on December 7, 1787—the first state to do so—and asserted that under the laws of this new nation, all people had inalienable rights. It continues to be a long journey to realize this ideal for all Americans, but it was an important first step toward true equality. . . . For national park aficionados, the biggest trivia question has always been "Which state does not have a unit of the National Park Service?" The answer: Delaware. Working closely with the Conservation Fund and the office of Vice President Joe Biden, we identified specific historical sites in Delaware that would be worthy of adding to the national park system. On March 25, 2013, President Obama designated the courthouse in New Castle and the historic properties of Woodlawn as the First State National Monument, part of the national park system.

Many spoke against the evils of slavery and some took matters into their own hands. Araminta Ross, born into slavery in 1822 on a Maryland plantation, took the name of Harriet when marrying John Tubman at age twenty-two. At age twenty-seven, Harriet Tubman slipped into the night and the dark waters of the eastern shore to follow the "drinking gourd" (the constellation known as the Big Dipper or Ursa Major) north to freedom. At great personal risk, she returned at least thirteen times to rescue other enslaved people and lead them north along the Underground Railroad. During the Civil War, she even went deep into the South as a Union spy and nurse to wounded soldiers, and she ran the contraband hospital at Fort Monroe for a time.

The State of Maryland was already working to build a new visitor center to honor the legacy of Harriet Tubman and nearby, still in agriculture, was the plantation where she had worked as a slave and then escaped in the dead of night. The Conservation Fund purchased the farm and, with Maryland State Parks, the NPS had a perfect partner to tell the story. On March 25, 2013, President Obama designated Harriet Tubman National Monument, on lands of the plantation where she was enslaved, part of the national park system.

The Underground Railroad has many routes, and one of those was along the Ohio River, well known as a center for abolitionism. One stop on that safe route was the home of Gabriel and Arminta Young, who escaped the racial violence of Kentucky in 1864 with their two-year-old son Charles. Charles Young grew up to enter the US Military Academy at West Point, the third African American to do so.

His distinguished military career took him around the world and into combat in the Philippines and Mexico. He rose to the rank of colonel. Among his many leadership roles, he commanded the famous Buffalo Soldiers, dispatched from the Presidio in California to protect the national parks of the Sierra mountains from poachers and developers. As the nation prepared for World War I, his experience and leadership should have led him to a promotion, but the military denied his advancement on false claims of poor health. To prove his physical condition, he rode his horse 500 miles to Washington, but clearly racial discrimination blocked his rise. He returned to Wilberforce University in Ohio and, in spite of the withering racism he had experienced, trained many young African Americans to follow his footsteps into the military.

Colonel Young's home was owned by the Omega Psi Phi fraternity and used by students at Wilberforce University. With their support and with funding through the National Park Foundation, the NPS acquired the home and grounds. On March 25, 2013, President Obama designated the Young home in Ohio as the Colonel Charles Young Buffalo Soldiers National Monument, part of the national park system.

Civil rights have been denied African Americans from the day they were brought here as enslaved people, but other ethnic groups also have suffered from discrimination in our society. This can be especially true during wartime. Soon after imperial Japan attacked the United States at Pearl Harbor and thrust the nation into World War II, President Franklin Roosevelt issued Executive Order 9066, requiring all persons of Japanese ethnicity on the West Coast to be rounded up and imprisoned for the duration of the war. Ten "confinement centers" were hastily built in remote places and over 100,000 people of Japanese ethnicity, many of whom were American citizens born in this country, were placed behind barbed wire and armed guards. Few places were as dreary as a bug-infested gulch on the Hawaiian island of Oahu known as Honouliuli. In 1988, President Ronald Reagan formally apologized to those families who were wrongly imprisoned for nothing other than their ethnicity.

The NPS already had an excellent relationship with the Japanese American Citizens League due to our stewardship of Manzanar, Minidoka, and Tule Lake Camps as existing park units. Garnering their support for the designation of Honouliuli was easy. Gaining access to the site, which was controlled by the Monsanto Corporation, was another story. On February 19, 2015, President Obama designated that gulch in Hawaii as Honouliuli National Monument, part of the national park system.

Following the Civil War, the northern workforce grew as former slaves migrated from the South. George Pullman of Chicago decided on a new business model, to build and lease fancy train cars that could be coupled to trains moving cross-country as we entered the twentieth century. Pullman staffed these cars with African Americans, especially the descendants of former slaves, as he thought they would be subservient to the Pullman car patrons. He trained the porters, paid them a living wage, and provided uniforms and a code of conduct.

Although still subject to discrimination and racism, these men developed pride in their work as porters, emphasized education in their children, and seeded the growth of a black middle class. In 1894, the porters participated in a major railroad strike that resulted in what we celebrate now as Labor Day. In 1925, they were organized by a young A. Phillip Randolph into the Brotherhood of Sleeping Car Porters Union.

The south side of Chicago is Barack Obama's stomping ground, where he canvassed door to door as a community organizer with the Developing Communities Project. Michelle Obama's grandfather was a Pullman porter. As a result, the Obamas were very receptive to the idea of creating a national monument to the Pullman porter legacy in their hometown. With their support, the National Park Foundation raised over $8 million to get the park started. On February 19, 2015, President Obama designated part of South Chicago as Pullman National Monument, part of the national park system.

When the crafters of our founding documents said "all men are created equal," they really did mean just men, especially white men who owned property. Fifty percent of the population, women, were excluded from voting, among other restrictions. Intent on changing this discrimination were two extraordinary women, Alice Paul and Alva Vanderbilt Belmont, who organized women to relentlessly protest in front of the White House and before Congress.

Paul founded the National Woman's Party (NWP) in 1916, the same year as the National Park Service was created. She led the first major battle for the Nineteenth Amendment to the Constitution, which would grant women the right to vote by ratification in 1920. In 1921, Belmont became the president of the party. Later, as the main benefactor of the NWP, she enabled the party to buy a historic home only blocks from the Capitol to use as headquarters. Working from this home since its purchase in 1929, over the next eight decades the National Woman's Party would craft the Equal Rights Amendment along with hundreds of pieces of legislation to recognize and protect the rights of women.

In the week leading up to the presidential proclamation that would establish this next national monument, I was sitting in the office of David Rubenstein, one of the cofounders of Carlyle, an investment firm worth billions. David had already been generous with his philanthropy to the NPS for major projects. He casually asked what I was working on and I told him about the NWP project. He said "need any money"? I said, sure, how about $1 million, knowing that the house needed repair but pulling the number out of my hat. David got up, walked out, was gone for a few minutes, came back and said "I'll cover that." On April 12, 2016, President Obama designated the home of the National Woman's Party as the Belmont-Paul Women's Equality National Monument, part of the national park system.

While the NPS had been working for decades on other civil rights sites, we had done little for the lesbian, gay, bisexual, transgender, and queer (LGBTQ)

Stonewall National Monument Designation event. Credit: National Park Service.

history, other than adding the gay rights activist Frank Kameny's house to the National Register of Historic Places in 2011. With the strong support of Secretary Sally Jewell, the NPS hosted a gathering at DOI of the leading LGBTQ scholars and historians. Sitting around my conference table, I asked, if the NPS could start with just one site, what would it be? The unanimous agreement was the Stonewall Inn in New York City.

While discrimination against people with visible differences such as race and gender has been present in our nation since the first colonists arrived, those in our society with a different sexual orientation have often suffered in silence, afraid to reveal their nature in public. There were some places they could gather and associate in relative safety, and one of those places was the Stonewall Inn on Christopher Street. The police nonetheless often raided the bar, arresting the patrons for drinking. On June 28, 1969, the Stonewall crowd fought back against the police. The riots spread to neighboring streets and word of their resistance spread around the nation. Soon LGBTQ rights groups formed in many cities, pivoting off the events at Stonewall to assert civil rights for the LGBTQ community.

The public meeting on May 10, 2016, in Greenwich Village was a fun gathering of activists and veterans of the movement. The only concern raised was that the NPS would make the story broad enough to include all those who were there and at the center of the protest, including transgender women of color. I assured the crowd our telling of the story would be fully inclusive. On June 24, 2016, President Obama designated the area around Christopher Street as Stonewall National Monument, part of the national park system. Not long after we established Stonewall, I took a call from a ranger who wanted to thank me. He had served, prior to coming to the NPS, in a suicide prevention call center in New York City. He said that establishing Stonewall as a unit of the NPS would actually save lives.

While the national park system does have sites that tell some of the civil rights history of African Americans, there are missing parts. The 1960s in America are often defined as the watershed of the civil rights movement, and there is no place more emblematic of the violence and the struggle than Birmingham, Alabama. Peaceful protests—even ones involving schoolchildren—organized effectively by civil rights leaders such as Dr. Martin Luther King Jr. were met with fire hoses, police dogs, beatings, and arrests. Dr. King used Room 30 in a hotel owned by African American businessman A. G. Gaston as his unofficial war room to plan each move, drawing national attention to the racial policies of this southern city. Bombing by segregationists became so common that the city became known as "Bombingham," but when a bomb set by Ku Klux Klan members at the Sixteenth Street Baptist Church killed four little girls on Mothers' Day, September 15, 1963, the nation said "enough." The momentum led to the passage of the Civil Rights Act of 1964.

Birmingham Mayor William Bell and his team as well as Congresswoman Terry Sewell were already doing a great job telling the tragic story of the civil rights movement in Alabama. With their extraordinary support, we found a way the NPS could add value to their educational efforts. On October 28, 2016, Secretary Jewell and I hosted a public meeting in the Sixteenth Street Baptist Church, sitting only feet from where the four little girls had been murdered fifty-three years earlier. On January 12, 2017, President Obama designated the A. G. Gaston Hotel and the area around the Sixteenth Street Baptist Church as Birmingham Civil Rights National Monument, part of the national park system.

The federal government during the 1960s was trying to exercise its authority to end the segregation and racial politics of Jim Crow in the South. One area where the federal courts had ruled against segregation was interstate transportation, which in those days meant the bus systems, predominantly Greyhound and Trailways. A group of white and black volunteers, known as the Freedom Riders, attempted to test this new ruling by boarding busses and riding together from the North to the South, stopping at southern bus stations and attempting to enter the segregated lunch counters and restrooms. The Freedom Riders arrived in Anniston, Alabama, on May 14,

1961, and when they tried to enter the "whites only" section of the station, they were violently attacked by members of the Ku Klux Klan. The tires on the bus slashed, it limped out of town and collapsed along the highway after less than five miles. The Klan attacked again, setting the bus on fire and attempting to burn those inside alive. They were able to escape but were again beaten until local law enforcement finally broke it up.

In October 2016, Sally Jewell and I boarded a bus parked in front of the bus station in Anniston, in the exact spot where the attack began in 1961. On the bus with us, besides local officials, were Hank Thomas, who had been on the Freedom Rider bus and been beaten, and Janie Forsythe McKinney, the then twelve-year-old girl who gave water to the Freedom Riders as they escaped the burning bus. Their stories were powerful and it was an honor to hear them in person. On January 12, 2017, President Obama designated the bus station and the highway bus burning site as Freedom Riders National Monument, part of the national park system.

With the infrastructure of the South in shambles after the Civil War and the country's leadership seeking reconciliation, the nation entered a period of history that is poorly understood and rarely interpreted: Reconstruction. Schools sprang up to quickly educate the newly freed African Americans, supported by churches in the North. Lands once held by plantation owners were subdivided and granted to many who had worked the lands enslaved, making them property owners. Laws were passed to ensure equality, and some African Americans, like Robert Smalls, were elected into public office. Violence against African Americans also rose and lynchings became a violent tool of oppression. Often called the "rehearsal for reconstruction," the town of Beaufort, South Carolina, and the nearby Penn Center demonstrated how the new South could be rebuilt. But political forces in the South soon overturned this effort, passing Jim Crow laws and making segregation the standard until the 1960s.

As we conducted the public meeting in Beaufort at the Darrah Hall and Brick Baptist Church in December 2016, I was reminded how little the public knows about the Reconstruction era, when there was an extraordinary opportunity to change the nation in the wake of the Civil War. In Beaufort, we recognized an opportunity for the nation to learn and heal from a very tragic period of our history. On January 12, 2017, President Obama designated sites at the Penn Center and Beaufort, South Carolina, as Reconstruction Era National Monument, part of the national park system.

The National Park Service tells stories through *place*, the physical location where history happened. These new national monuments are not only the story solely of African Americans; Japanese Americans; women; or people who are lesbian, gay, bisexual, or transgender, they are *American* stories. By preserving these places along with others like Gettysburg, the Statue of Liberty, and the Grand Canyon, we are honoring their contribution to a nation that is seeking to achieve the ideals articulated in the Constitution and the Declaration of Independence. Threads through each bind them together: A. Phillip Randolph learned his labor activist skills organizing porters at Pullman and brought these skills to Birmingham. Dr. Martin Luther King Jr. often retreated to a small cottage at the Penn Center, since the days of Reconstruction still a safe place to gather. Harriet Tubman worked in Beaufort as a Union spy, and the Ohio home of Colonel Charles Young had a secret tunnel that linked to the Underground Railroad. César Chávez drew inspiration from the peaceful protests of Dr. King. Each group that fought for its rights—farm workers, guides to the Underground Railroad, Japanese Americans, the Suffragettes, the LGBTQ crowd at the Stonewall Inn, the peaceful marchers in Birmingham, and the Freedom Riders—all took lessons and inspiration from each other. And the Pullman porters set in motion the rise of the black middle class, whose descendants include First Lady Michelle Obama and whose husband became our forty-fourth president and established these new parts of the national park system so that we can tell a more complete story of America. By visiting these places and learning their stories, we can draw inspiration from their bravery and sacrifice to continue the fight for civil rights for all residents of this great nation.

While these twelve new park units broadly increase the diversity of the national park system, I was also pursuing other sites of importance, some through presidential proclamation and some through legislation. One of the more complex sites was the Manhattan Project, with three sites that recognize the development of the atomic bomb during World War II. With Oakridge, Tennessee, Los Alamos, New Mexico, and Hanford, Washington, all under study, our historic preservation colleagues in Japan were very concerned that the designation of these historic sites would be construed as a "celebration" of the use of the atomic bomb on Hiroshima and Nagasaki, Japan. I was interviewed by the Japanese media on several occasions to reassure them that this was not a celebration but a historical reminder intended to deter the future use of atomic weapons. The Manhattan Project also was a way to give greater emphasis to the women who worked on the project. The Manhattan Project National Historical Park became a unit of the NPS on November 11, 2015, by act of Congress.

Over the course of my tenure I was able to add Valles Caldera National Preserve, New Mexico, to the NPS system. This spectacular high mountain forest and meadows, formed by a collapsed caldera, should have been part of the NPS long ago, but when it was purchased from private owners in 2000 for $101 million, the Republican-led Congress tried to set it up as a financially independent "preserve" under the management of the US Forest Service. Operating funds were to be generated through hunting permits and timber sales. It soon became a private hunting ground for the wealthy and remained off limits to the general public. This ill-conceived model failed, and Congress recognized that the NPS would be a much better manager and legislatively added it to the NPS in December of 2014.

For decades, the idea of a national park in the north woods of Maine had been suggested, generating a great deal of controversy. Northern Maine is composed of millions of acres of private timber lands, and the population's livelihood has been linked to forest, sawmill, and paper mill jobs for generations. But the economy was changing and the big timber companies were selling off their land to private developers. The last mill in the area had closed and was sold for scrap and the formerly robust towns were struggling. Over a period of twenty years, Roxanne Quimby, the founder of Burt's Bees, purchased lands from willing sellers along the border with Baxter State Park with the publicly stated intent to donate them to the National Park Service to create a park. Local opposition to the idea was strong, particularly around Millinocket, which had once been a successful mill town and now had only one remaining open storefront in the entire downtown. The community leaders longed for the return of the timber industry and did not see value or opportunity in becoming a gateway to a national park. In contrast was Acadia National Park, only a few hours' drive away and with a sustainable tourism economy in the gateway town of Bar Harbor. Roxanne had offered her lands as a gift to the NPS in 2008, but NPS Director Mary Bomar, serving under President George W. Bush, rebuffed the offer, stating the NPS was not interested in adding new parks.

When I became director in 2009, I reached out to Roxanne and began building the political capital to make a park in Maine a reality. As well-meaning as Roxanne was, her blunt style of speaking infuriated the locals and the congressional delegation, and was making an already difficult task even more challenging. Secretary Salazar and I convinced Roxanne that in order to win more support, she needed to turn the park effort over to her son, Lucas St. Clair. Lucas proceeded to meet with anyone and everyone, drinking, as he said, "a thousand cups of coffee." Born and raised in the Maine woods, the charismatic Lucas built trust in the fiercely independent Mainers, suggesting the park was the only viable future for jobs, the economy, and for their outdoor lifestyle. I spent several hours with Senator Susan Collins discussing what the NPS would bring to the region, and I traveled to listen to the communities with Senator Angus King. In May of 2016, Senator King and I hosted a public meeting at the University of Maine, Orono, and took testimony for three hours. Opposition was still strong but the room was pretty evenly divided. As each person spoke, I wrote down their first name and made a few notes. After all who wanted to speak had their opportunity, I stood up and called out each of the speakers by their first name and responded directly to their concerns. While it may not have changed their minds, every person knew they had been heard. Although we really wanted this to be done legislatively, time was running out. We turned to the president for a proclamation and set up a process for Roxanne to donate 83,000 acres of north Maine woods to the National Park Service along with her $20 million endowment to the National Park Foundation. The resulting proclamation established Katahdin Woods and Waters National Monument. Quimby's generous donation—made on August 25, 2016, the Centennial of the NPS—was the largest land gift to the NPS in its history, exceeding that of the Rockefeller family's lands donated in Grand Teton, Acadia, and the Virgin Islands National Parks.

Then, of course, there were some efforts to add new parks that just did not happen, because of politics, lack of local support, or timing that was just off. I worked hard to get the North Shore Cliffs of Molokai added to Kalaupapa National Historical Park by donation from a wealthy philanthropist, but resistance from the local Hawaiians and the State of Hawaii made it impossible. Eminent biologist E. O. Wilson encouraged our designation of the Mobile Tensaw River in order to protect its extraordinary biodiversity, but the conservative politics of Alabama has so far prevented any real traction. At the end of my tenure as director, twenty-three new units had been added to the system, broadening the mission across the landscape and the sweep of history.

As Our Nation Evolves, So Too Must the National Park System

In order to serve the American public to its fullest potential, to help the nation gain an authentic presentation of our complex history, both the good and the bad, and to preserve the very best of our natural resources and landscapes and our most important cultural heritage sites, the NPS needs to be free to grow. The professional evaluation of potential new parks should not be subject to political control. In 1978, the NPS was authorized to evaluate and recommend annually new additions to the national park system. In 1998, that authority was removed by a Republican-led Congress, and the NPS is now only allowed to study new parks when specifically authorized by Congress. Even when the NPS is authorized to study an area for potential designation and it is found to meet the criteria, if the administration at the time is hostile to expansion, the NPS is forced by the Department of the Interior and OMB to testify against its own recommendations. During the administration of President Ronald Reagan, the NPS testified against every new national park unit proposal before Congress, and no new monuments were added. These experiences, repeated over and over, are extraordinarily frustrating to the professionals in the NPS, who worked with communities, advocacy groups, and members of Congress to complete the required study and then, in a congressional hearing, must oppose the designation.

As director, Jonathan commissioned a new National Park System Plan to outline the gaps in the existing system by broad categories,[17] recognizing there are still omissions, in particular related to the contributions of women and people of color. Moreover, initiatives in recent years, such as Dr. E. O. Wilson's "Half Earth" proposal, which seeks preservation of half of the lands and waters of the world for our survival's sake, cannot be fulfilled without significant new additions to the national park system, America's premier preservation agency.

Our combined ninety years of experience lead us to conclude that growth of the national park system—to include all of the American story and the very best of its natural heritage—can only be achieved by removing the National Park Service from the Department of the Interior and establishing it as an independent agency, where its professional, scholarly, scientific, and historically based findings can be fully described to the public, the president, and the Congress.

TWO

ALASKA: DOING IT RIGHT THE FIRST TIME

For two boys who grew up in rural Virginia hiking, hunting, and fishing within the woods and waters of the Blue Ridge Mountains, Alaska was just a dream, a very far-off place. While traveling thirty miles to another creek or valley in Virginia was an adventure, we vicariously visited wild country through the writings of Zane Grey, Sigurd Olson, and Wallace Stegner. We traveled west with Ward Bond on "Wagon Train" and experienced nature up close with Walt Disney. But Alaska called to us as the ultimate adventure. Little did we know that, one day, we would have the opportunity to not only experience but influence the protection of the wildest place left in the United States.

An early and short-sighted mistake made by Congresses and administrations regarding new parks was that most unit boundaries were set in law along straight section lines. This began with Yellowstone, whose western boundary can be clearly seen from satellites that show US Forest Service clear-cuts right up to the line. As the NPS and its conservation constituencies began to consider new parks in Alaska, a new approach to park boundaries was widely agreed upon: that is, boundaries would be drawn on geographic features to incorporate whole ecosystems and watersheds as an expression of ecological consciousness and improved wildlife management. Thus, "doing it right the first time" became the mantra for new Alaskan parks and refuges, enabling agencies to avoid going back to Congress for multiple boundary changes in future years.

Alaska—Seward's Folly, as it was called when it was bought from Russia in 1867—then encompassed 375 million acres, virtually all federal lands, including the most pristine wild landscapes in North America. At the time that Alaska was granted statehood by act of Congress in 1959, the new state was entitled to select 100 million acres of these federal lands for its own.

In January 1965 NPS Director George Hartzog appointed an internal NPS Alaska Task Force to prepare maps and recommendations for future parks to be taken from the land remaining with the federal government. The report, *Operation Great Land*, finalized in 1969, proposed 76 million acres of new parkland in Alaska.

At President Lyndon Johnson's request, Secretary of the Interior Stewart Udall had the NPS prepare a number of national monument proclamations to consider in Alaska. But then, just days before he left office, Johnson declined to sign the proclamations in a fit of pique at Udall for having announced the likelihood of the monuments before the president acted. Udall then issued Secretarial Order 4582 on January 17, 1969, only a few days before president-elect Nixon was sworn into office, which effectively froze state land transfers pending resolution of native land claims and also temporarily halted the Prudhoe Bay oil pipeline construction.

DESTRY

In 1972, Hartzog assigned a key member of his internal Alaska Task Force, Ted Swem, to lead the NPS Alaska park studies team, and NPS career park planners John Kauffmann, John Reynolds, and several others moved to Anchorage.

Kauffmann became a good friend of mine during these planning years, a friendship that continued after he retired from the NPS and became a board member of the NPCA and a founder of Friends of Acadia, another group with which I worked closely. Kauffmann is also highlighted in perhaps the most well-known book about Alaska, *Coming into the Country* by John McPhee, published in 1977. This book became something of a rallying cry, likely read by every grassroots activist working with the Alaska Coalition, the main advocacy NGO working to define these public lands during the subsequent legislative years. Kauffmann did not need an NPS career, as he was the wealthy son of the owner of the *Washington Star* newspaper, but his deep devotion to wilderness and efforts to preserve it, especially in the Brooks Range of Alaska, propelled him into his career.

By the time Congress enacted the Alaska Native Claims Settlement Act of 1971 (ANCSA), Alaska had still not secured final selections of federal land for transfer to state ownership, largely because officials were waiting on completion of mineral and timber surveys in order to select the most commercially valuable lands. ANCSA effectively froze finalization of most state land selections until native claims, encompassing 44 million acres of land, were settled.

Section 17d(2) of ANCSA required the Secretary of the Interior to select for permanent retention by the federal government up to 80 million acres of land suitable for national parks, wildlife refuges, and national forests. The fact that this provision was included in ANCSA at all was largely due to efforts by NPS Director Hartzog and Senator Alan Bible (D-NV), chair of the Subcommittee on National Parks.

In 1971, Hartzog arranged to take Bible on a tour of the proposed park sites in Alaska, for which they were accompanied by Dr. Ed Wayburn, president of the Sierra Club, and Celia Hunter, owner of Camp Denali and chair of the Wilderness Society Board.[1] Both Wayburn and Hunter later became good friends of mine, and both played critical roles in the work of the Alaska Coalition to secure enactment of the Alaska Lands Act. Senator Bible came away from his Alaska trip convinced not only that these federal lands deserved preservation as national parks but also that their integrity was threatened by state and native selections if not addressed in legislation.

During the ANCSA debate, it was Senator Bible's amendment that became section 17d(2). Bible sought the amendment with the intent that it secure the NPS *Operation Great Land* report's recommendation of 76 million acres for new parks. As things turned out, due to pressure from the Alaska delegation and governor, all the other federal land management agencies were also included in the amendment.

As the Nixon administration's work began in early 1972 to develop the d(2) park recommendations, Hartzog's Alaska planning team, led by Ted Swem, needed to develop the necessary orders for Secretary of the Interior Rogers C. B. Morton to withdraw from state or native selections the lands recommended for parks and other conservation system designations. The team in Anchorage was divided into five groups, each with a geographic assignment to compile all possible data on each proposed park's national significance and outstanding values. John Kauffmann led the team that produced the plans for Gates of the Arctic, Noatak, and Cape Krusenstern; John Reynolds was team captain for the sites that became Bering Land Bridge, Kobuk Valley, and the expansion of Denali. The core team included Dick Stenmark, who likely had the deepest direct on-the-ground knowledge of the sites being proposed for parks of any NPS employee. I met with Dick many times over the next nine years and was always impressed by his professionalism and deep knowledge of Alaska.

President Nixon fired NPS Director Hartzog in December 1972.[2] Fortunately, given the appointment of an unqualified political as NPS director, Assistant Secretary Nat Reed and his staff took firm charge of NPS policy and decisions, including continuation of the Alaska park studies. Reed quickly established the Alaska Planning Group, managed out of his office and chaired by Ted Swem. This team covered the interests of all three agencies under Reed's management: the NPS, the Fish and Wildlife Service, and the Bureau of Outdoor Recreation, the agency doing the river and trail studies. For the next year, the Alaska Planning Group was the key to developing the administration's final recommendations for Alaska park and refuge legislation.

As the representative of the National Parks Conservation Association (NPCA), I was involved with the important meeting in December 1973 when Secretary Morton announced his decision on the withdrawal of 80 million acres of proposed parks and refuges, an action that had the effect of protecting these lands until such time as Congress passed authorizing legislation for these new parks and refuges. I had lobbied hard, along with most other national conservation organizations' representatives, to convince Secretary Morton that the new national parks and wildlife refuges being planned for Alaska warranted his support.

Unfortunately, from a conservation perspective, the Morton recommendations to Congress proposed 19 million acres of new National Forests in the interior of Alaska, multiple-use lands open to timber harvest and mining that were opposed by NPCA and all other national conservation organizations. Worse for the parks and refuges, the proposed Morton boundary lines were drawn on real estate township section lines rather than following natural geographic features, which are more readily identified on the ground and which tie directly to ecosystems and native species habitats. The NPS had already lost considerable acreage around the park boundaries in political deference to pressures from the Alaska delegation and due to Nixon administration political infighting that favored more federal lands left open to multiple-use development.

To be fair, Morton's assistant secretary for Fish and Wildlife and Parks, Nathaniel P. Reed, and his two deputies, Doug Wheeler and Curtis "Buff" Bohlen, were, in my opinion, the very best conservation-oriented officials ever to serve in a Republican administration. In fact, Buff, a consummate professional, came back to work at DOI in both the Carter and Clinton administrations, where I worked with him frequently. All three continued in distinguished conservation careers long after they left federal service. I worked closely with all three of them during their years at DOI and since. I have had an especially close working relationship with Doug in nearly all his subsequent jobs—as CEO of the Conservation Foundation, vice president at World Wildlife Fund, founder of American Farmland Trust, secretary of the Resources

Agency for the State of California, executive director of the Sierra Club, and chair of the Chesapeake Conservancy. Nevertheless, under relentless pressure from the Alaska congressional delegation, parks and refuge proposals were whittled down considerably in the final Morton legislative proposal.

One additional regulatory complexity that the NPS faced for its new park proposals was the National Environmental Policy Act (NEPA), which required federal agencies to prepare an environmental impact statement (EIS) when their proposed actions could "significantly affect the quality of the human environment." John Reynolds became team captain for the EIS preparation. NEPA was a new law, which up to that time NPS leadership thought should not apply to it since they were the "good guys" and thought NEPA was for the "bad guys." Nevertheless, John and the other planners prepared an exemplary set of EISs that fully documented the resources and values of the proposed parks.

Another critical event to greatly influence the outcome of the legislation was an Alaska study trip that Ted Swem arranged for House Subcommittee Chair John Seiberling, Representative Goodloe Byron, and his key subcommittee staffer Cleve Pinnix to tour the proposed park sites in the summer of 1976. Following their firsthand experiences on this trip, both Seiberling and Byron became strong advocates for the new national park proposals in Alaska. I worked closely with both of them on the ensuing legislation. Tragically, Goodloe would die of a heart attack in late 1978, but his widow, Beverly, won his seat and also became a strong supporter of the Alaska Coalition.

As the new Congress convened in 1977, Seiberling became chair of a newly established subcommittee focused fully on the Alaska lands bill. These d(2) lands then became the subject of the most intensive and extensive land conservation legislative lobbying campaign in American history.

In early 1977, just as I and my conservation colleagues were organizing the Alaska Coalition to include all the major national conservation organizations and many others, President Jimmy Carter appointed former Idaho Governor Cecil Andrus as Secretary of the Interior and brought in Robert Herbst, former Natural Resources Secretary in Minnesota, as assistant secretary for Fish and Wildlife and Parks.

Herbst had a strong bias in favor of US Fish and Wildlife Service (USFWS) refuges over NPS parks, because national park status eliminates sport hunting, and thus he prepared a proposal to convert most of the park proposals to wildlife refuge proposals. The USFWS especially wanted the Northwest Alaska proposed park sites of Noatak, Kobuk, and Bering Land Bridge. Although these refuge proposals did not survive internal administration decisions, Congressman John Dingell introduced an Alaska lands bill with only wildlife refuges—no national parks—which fortunately was never seriously considered by Congress.

To Andrus's credit, he quickly brought in Cynthia Wilson from the National Audubon Society to be his personal assistant for Alaska lands. A friend with whom I had worked on numerous conservation bills in the past, she was highly skilled, articulate, and driven to succeed. Hers was the largely thankless task of getting consensus among all the agencies with conflicting Alaska lands agendas.

At that time, individual DOI agencies were allowed to file their own legislative reports to Congress on pending bills that would directly affect them. This long-standing practice later came to an abrupt end in 1981 when President Reagan's DOI Secretary James Watt stopped it, and it has not been allowed by subsequent political appointees in DOI since. NPS Director Bill Whalen, who was selected as director in July 1977, assigned his Alaska Lands Task Force the job of writing a report on HR 39. HR 39 was the House bill for the conservation version of the Alaska National Interest Lands Conservation Act (ANILCA), introduced by Chair Morris Udall and Subcommittee Chair Seiberling and others early in 1977. The number 39 itself became a famous grassroots rallying cry and was despised by the Alaska delegation. Udall had to fight hard to retain that number for the reintroduction of the conservation bill when the next Congress convened in 1979.

Work on the NPS report on HR 39 largely fell to Jim Pepper, a smart and savvy career NPS legislative specialist who was detailed to the Secretary's office to work on Alaska issues. The report was submitted in August 1977. In addition to all of the original park proposals from the 1969 report, the NPS proposed "instant" wilderness (that is, established directly in the bill without need for subsequent study and separate legislation) for millions of acres of these new parks, introduced the idea of "national preserves" to allow hunting in areas around national parks, and proposed two new national parks to be carved out of the existing Tongass National Forest in southeast Alaska, for Admiralty Island and Misty Fjords.

From 1976 to 1980, the vast majority of national conservation organizations, including the NPCA, subsumed their individual agendas and inherent competition to merge their staffs, energies, and money into the Alaska Coalition in order to more effectively advocate for maximum conservation designation of federal lands in Alaska. This action remains today the only instance of this level and extent of interorganization conservation cooperation.[3]

For our four-year campaign, I was one of the six members of the Alaska Coalition Steering Committee, representing the National Parks Conservation Association, along with the Sierra Club's Chuck Clusen, Friends of the

Earth's Cathy Smith (and later Pam Rich), the Wilderness Society's Doug Scott, and the National Audubon Society's Steve Young; Clusen and Smith were the cochairs. To this group were added a rotating individual from one of several small conservation organizations based in Alaska, initially Paul Peyton from the Southeast Alaska Conservation Council, and Dee Frankforth, an environmental activist from Anchorage, Alaska. Overall, the coalition was composed of several hundred national, state, and local nonprofit organizations. We benefited greatly throughout the campaign from the wise counsel of several "elders of the tribe," Mardy Murie, Dr. Ed Wayburn, Celia Hunter, and Bob Cahn.

During this time, beyond my lead advocacy for the proposed national parks, my main coalition role (other than serving on the policy steering committee) was as chief Senate lobbyist; others focused on the House. The steering committee met weekly and the legislative lobbying team met daily for updates and strategy discussion.

The single most important coalition strategy was an unprecedented level of grassroots organizing and activating across all fifty states. Such grassroots engagement was critical, because the normal congressional process would have deferred to the wishes of Alaska's congressional delegation when only that state was affected by pending legislation. In this instance, however, the coalition made the public case that the fate of these federal lands, owned by all Americans, should be decided in the national interest, and not by the three members of the Alaska delegation.

This grassroots strategy was aided immeasurably by the complementary strategy of House subcommittee Chair Seiberling. Over the course of three years, he held dozens of field hearings on the pending Alaska Lands Act in states all across the country, where local activists, organized by the coalition, could testify in person and garner positive media coverage for the bill. These meetings made the highly visible and effective point that these lands' fate warranted all Americans to take a stand and force a legislative outcome based on the national interest.

Alaska then, and even to some extent today, was a relatively unknown landscape outside its few developed cities and towns. Very few people could have been said to be experts on that unsettled landscape proposed for parks and refuges. Two NPS professionals who were, were Bob Belous and Dick Stenmark. Bob worked mostly out of the NPS Alaska regional office in Anchorage but roamed the backcountry with his cameras, notebooks, and maps. Dick moved from Anchorage to Washington, DC, and, in 1979 and 1980, was duty-stationed there with the Senate Energy and Natural Resources Committee specifically to prepare the maps to correspond to the bill language, which changed many times through debates and compromises. I met with them both numerous times over these four years, and it is fair to say that their expertise ought to be credited with holding onto the main goal for park boundaries in Alaska: to get it right the first time, that is, to place boundary lines along geographic points on the landscape and not on the straight lines of township borders, so as to incorporate whole watersheds, river valleys, ridge lines, and other natural features within the designated unit. These are the boundaries that best preserve the biological and geological integrity of the parks.

Such a bold and unusual coalition strategy was necessitated by the very strong opposition to the conservation objectives inherent in the proposed legislation. Both Alaska senators, Ted Stevens and Mike Gravel, and the lone House member, Don Young, as well as the Alaska governor, Jay Hammond, and their oil, gas, mining, and timber industry allies were all adamantly opposed to HR 39.

Although the goals of the coalition were strongly supported by key leadership in the House—especially Udall and Seiberling—Representative Young sought a much weaker bill. He proposed millions fewer acres for national parks, wildlife refuges, and wilderness, millions of acres that would instead have been placed under the US Forest Service and Bureau of Land Management for multiple-use and resource extraction purposes.

In May 1978, HR 39 reached the floor of the House for a vote. Representative Young offered the first amendment to test support, or lack thereof, by proposing to delete 4.5 million areas from parks and refuge designations, substituting multiple-use management by the BLM and USFS. That amendment was defeated 251–141, and HR 39 went on to final passage by a margin of 279–131.

The Senate, however, was a very different story, with Committee Chair Henry M. "Scoop" Jackson seeking to forge a compromise between the Alaska delegation's demands and the House leadership and coalition's position. These differences boiled down to fights over 1) acreage to be designated as national parks versus national preserves (to allow hunting in the latter); 2) acreage to be designated as national wildlife refuges versus left for state selection or managed by the BLM (and so not part of the federal conservation systems); and 3) acreage of parks and refuges to be overlain with wilderness designation (or not). To Jackson's credit, from the outset he did not support the Alaska delegation position, which sought millions of acres of national forest (multiple use, timber cutting, and mining) in the interior of the state.

In 1977–78, our main Senate champion was New Hampshire Senator John Durkin, with whom I worked closely to develop strengthening amendments to the version of the bill being supported by Chair Jackson, and to vigorously oppose weakening amendments offered by Senator Stevens. Durkin offered our suggested amendments during the Energy and Natural Resources

Committee markup (amending and voting) sessions on the bill, but it was an uphill battle.

After turning down both of Durkin's strengthening amendments and much weaker ones offered by Republicans on the committee, an Alaska lands bill, still much weaker than HR 39, was voted out of the Senate committee in October 1978, just eight days before Congress was scheduled to adjourn for the year. Desperate to get a bill finalized, Senator Jackson and Chair Udall convened several closed-door informal meetings with Seiberling and a few other key legislators to seek a compromise that could pass both Senate and House before Congress ended. Coalition members were, of course, excluded from these secret negotiations, since they deviated from the established procedure where the House and Senate would conference only with bills that had already passed both bodies.

In the end, Jackson and Udall agreed to a bill text that sought to compromise their differences, but Alaska Senator Gravel opposed it and announced that he would filibuster it on the Senate floor. This forced Senate Majority Leader Robert Byrd to pull the bill altogether just prior to adjournment of Congress, thus ending legislative action without a bill in the 95th Congress. Although gloating over his perceived victory, Senator Gravel feared that President Carter would take administrative action to protect these federal lands in Alaska. He was heard to say he felt that President Carter did not have the guts to use the Antiquities Act to proclaim these lands as national monuments. Fortunately, events proved him wrong.

Throughout his four-year term, President Carter and Interior Secretary Andrus had made getting an Alaska Lands Act passed their highest environmental priority. We met with Andrus and his senior staff often, and I joined a dozen of my conservation colleagues for a summit meeting with President Carter at the White House in the summer of 1978.

We urged President Carter and his administration, if Congress failed to pass the Alaska Lands Act, to take immediate action. Our concern was to forestall efforts by the state and Alaska delegation to block future conservation designations, which the president could do by invoking his authority to proclaim national monuments on federal lands at his sole discretion.

A critical deadline of December 18, 1978, for legislation on the d(2) lands had been set in the Alaska Native Claims Settlement Act back in 1971. If no d(2) bill was enacted by that date, federal land selection restrictions would be lifted for native and state land selections, which if done would likely then foreclose subsequent conservation land designations by Congress. President Carter and Secretary Andrus had warned Congress earlier in the year that if no Alaska lands bill was enacted, the administration would take direct action.

*To Destry Jarvis
With thanks for your support!*   *Jimmy Carter*

Destry Jarvis at Alaska Coalition 1978 meeting with President Jimmy Carter. Credit: National Archives, Carter Presidential Library.

On November 16, Secretary Andrus used emergency authority to withdraw 110 million acres of federal lands in Alaska from mining claims and state selection for another three years. On December 1, President Carter used the authority of the 1906 Antiquities Act to proclaim 56 million acres as national monuments, accomplishing permanent preservation unless a subsequent Congress enacted a bill altering these designations.

Senator John Durkin was defeated for reelection in the fall of 1978, and Massachusetts Congressman Paul Tsongas was elected to the Senate, becoming the main Senate champion for the coalition's Alaska Lands Act in 1979–80.

When the new Congress convened in early 1979, all sides wanted to get an Alaska Lands Act passed, just not the same bill. The Carter administration, the Alaska Coalition, and House Committee Chairs Udall and Seiberling along with Senate coalition champion Paul Tsongas (among many other members) were aligned on one side, while the governor of Alaska, the three-member Alaska congressional delegation, and major energy, mining, and timber companies, among others, were aligned on the other. At that point,

Alaska's governor appointed an industry-funded group, Citizens for the Management of Alaska Lands (CMAL), to carry on the lobbying campaign in Congress to stop HR 39 or to substitute a multiple-use bill.

During this time, I joined the coalition's lobbying team for regular, often daily, meetings with senior administration and congressional leaders, including Secretary Andrus, Chair Seiberling, and Senator Tsongas, to plan strategy and discuss potential amendments.

On January 15, 1979, House Committee Chair Udall quickly introduced a new version of HR 39. It now included subsistence provisions that brought active Native Alaskan support to the bill, because the natives were guaranteed rights to continue traditional hunting, trapping, and fishing in these new conservation areas. However, a huge lobbying effort by the state, the oil and gas industry, and the National Rifle Association resulted in HR 39 losing in committee by one vote. Undeterred, with a cranked-up coalition grassroots lobbying campaign that especially sought Republican support, Udall and Republican John Anderson (IL) introduced a substitute amendment on the floor of the House, which passed 268–157. The May 16, 1979, vote on final House passage of HR 39 was an overwhelming 360–65.

Throughout 1979 and into early 1980, direct lobbying in the Senate by coalition members, including myself, and often accompanied by influential board members from the coalition organizations, intensified in an effort to sway votes. One of my most memorable lobby days was escorting an NPCA board member, Mrs. W. L. Lyons (Sally) Brown, to see her home state senator, Wendell Ford (D-KY), a member of the Energy and Natural Resources Committee, where markup on the bill was occurring. Senator Ford was hesitant to support the coalition's strengthening amendments, preferring Chair Scoop Jackson's middle-ground compromises. At one point in the tense meeting, Sally shook her finger in the senator's face and said, "My family controls one-fifth of the economy of Kentucky, and you will do what I say . . ."[4] Senator Ford thereafter supported most of the coalition's amendments, though we still did not have enough votes to prevail in the committee on the final vote to report the bill.

Chair Jackson, in a highly unusual move, allowed Alaska Senator Stevens to fully participate in committee hearings. Stevens questioned witnesses, took part in debates and more than a dozen multihour committee markup sessions, and offered his own amendments, even though he was not a member of that committee (he did not get to vote in committee). Largely as a consequence of these maneuvers, committee action on the bill dragged on throughout 1979. Senate committee markup did not begin until early October 1979, and the committee reported its weak version of the bill on October 30 by a vote of 17–1, with Senator Tsongas, the coalition's champion, casting the lone vote against.

This committee bill would have taken 40 million acres of conservation lands out of the Alaska Lands Act. By early November, Tsongas, Senator William Roth (R-DE), and twelve cosponsors announced their intent to offer a far stronger conservation bill—one supported by the coalition—during the full Senate floor debate. Senator Gravel announced his intent to filibuster any bill. With no time remaining in the session, leadership pulled the bill from the floor and carried it over until Congress reconvened in early 1980. At that time, Senator Stevens announced that he had negotiated a five-month delay with Senate leadership before the Alaska Lands Act could be brought up on the floor, including an agreement that fourteen amendments could be offered, but only five by Tsongas, with nearly fifty hours of debate scheduled. This unusual maneuver threatened to delay final passage until after the fall election, pushing it into a lame duck session in which uncertainty is the rule.

As a consequence, in February 1980 Secretary Andrus signed an order, using his administrative authority under section 204 of the Federal Lands Policy and Management Act, permanently withdrawing another 40 million acres from state selection or mineral entry. These were lands that HR 39 would have designated as national wildlife refuges, and Andrus's action again pressured Congress to act on the bill.

When the bill finally came to the Senate floor in late July 1980, a series of test votes confirmed that the coalition had the votes to prevail. A weakening amendment lost by a vote of 66–30. Senators Jackson and Hatfield then offered a substitute bill, endorsed by Stevens and Gravel, which lost by a vote of 62–33. It was then clear that the coalition and its Senate allies likely had the votes to prevail on the whole bill. However, Stevens, a master of senate rules and procedures, announced his intent to offer eighteen secondary amendments. This tactic led Senate leadership to again pull the bill from the floor and direct committee leaders, including Tsongas and Stevens, to craft a compromise to the committee's reported bill—one that could pass.

That bill, known as the Tsongas-Jackson-Roth-Hatfield substitute, was brought back to the Senate floor on August 4. This version of the bill was nearly the same as HR 39 for NPS lands, though with more preserves, but deleted 26 million acres of wildlife refuge lands. At that point, Gravel again launched into a filibuster, causing endless delays. Senate leadership then filed a cloture petition to cut off debate. Thereafter, the substitute passed, and the Senate passed the full bill by a vote of 78–14.

House leadership had not been consulted on the substitute compromise, but with the 96th Congress rapidly coming to an end, the election of Ronald Reagan as president, and the Senate majority shifting to Republicans, the House reluctantly accepted the Senate bill with no changes. It passed on November 12 and was sent to President Carter, who signed it at a White House ceremony, which I attended, on December 2, 1980, as the Alaska National Interest Lands Conservation Act (ANILCA), Public Law 96-487.

Enactment of ANILCA doubled the acreage of both the national park system and the national wildlife refuge system. It was made possible by the most expansive and unified lobbying campaign ever undertaken by the American conservation movement as well as by a highly supportive president, Secretary of the Interior, and a willing Congress with Democrat majorities in both House and Senate.

After nine years of debate and legislative action, ANILCA added 43.6 million acres to the national park system, consisting of ten entirely new units and significant additions to three established parks, Denali, Glacier Bay, and Katmai. This was some 30 million fewer acres than NPS Director Hartzog had proposed in 1965.

In the final analysis, ANILCA was a compromise. It included far fewer wilderness designations than the coalition had supported, and it had various management provisions that would prove to be problematic in the years ahead. The coastal plain of the Arctic National Wildlife Refuge, summer calving ground of Alaska's largest caribou herd, was left out of wilderness due to the possible presence of oil and gas in productive quantities; a future road corridor was carved out across the south side of Gates of the Arctic National Park to connect to the Prudhoe Bay Trans-Alaska Pipeline Haul-Road to allow for future copper mining in the Ambler District;[5] and the subsistence provisions of the bill were amended to extend these rights beyond Alaskan Natives to anyone living in rural Alaska.

While the ANILCA signing ceremony was largely a celebration of a great achievement for conservation, two leaders, with diametrically opposed views, Senator Stevens and the Sierra Club's Dr. Ed Wayburn, were both quoted that day about the future of Alaska lands. Stevens, ever belligerent, stood at the podium and said, "We are not finished, Mr. President . . . we've really just started. We know that the time will come when those (protected) resources will be demanded by other Americans." With different intent but similar words, Ed said, "The Act is not an end, but a beginning. Alaska's superb wildlands must have more secure protection. And all concerned Americans will continue to work together until we gain it."

As 1981 began, I was elected chair of the Alaska Coalition. My task was to lead the effort to prevent the new Congress from passing ANILCA amendments that were supported by the Reagan administration over NPS objections, to weaken the conservation achievements of that landmark statute.

Senator Stevens, now in the majority and an official member of the Energy and Natural Resources Committee, soon introduced a bill, S 2826, to convert millions of acres of the new Alaskan national parks to national preserves, to allow sport hunting and commercial trapping. Representative Don Young introduced the companion bill, HR 6977. Mostly because the majority in the House was still controlled by the Democrats, who would not consider opening up ANILCA to amendments so soon after enactment, neither bill passed.

Of course, the election of Ronald Reagan as president and his subsequent appointment of James Watt as Secretary of the Interior early in 1981 resulted in huge problems for the NPS in implementation of management for the ANILCA parks and preserves. In fact, Representative Don Young's top legislative aide, Bill Horn, who had done most of the House staff work in opposition to ANILCA, was appointed to key jobs in the DOI. First he became associate solicitor for Conservation and Wildlife, with lead responsibility for drafting regulations to implement ANILCA, and later he was appointed assistant secretary for Fish and Wildlife and Parks, with direct day-to-day management of both the NPS and the US Fish and Wildlife Service.

JONATHAN

As always the case, after Congress acts to establish a new park, the NPS is charged with enacting both the letter and the intent of the legislation, which are sometimes not the same thing. In the case of ANILCA, this was particularly complicated and I was lucky enough to be directly involved in figuring it out. Immediately after President Carter established the new national monuments in Alaska in 1978 under the Antiquities Act, that designation made hunting illegal in the areas managed by the National Park Service. While hunting in the lower forty-eight states is a sport, a pastime, and a hobby, in Alaska it's a lifestyle, part of the culture, and, for some, it supplies their primary source of food. This change was not received well in the bush of Alaska, nor were the rangers the NPS sent to enforce the new restrictions. As much as I wanted to be, I was not in the first wave of national park rangers who were dispatched to bring some sense of management to the new areas of Alaska. In 1980, I was offered a ranger position in Kotzebue, but when I suggested this to my wife Paula, she said "Be sure to write." It took fourteen more years to convince her that this would be fun!

Those first rangers, sent to enforce Carter's proclamation, were not greeted warmly by the local residents, who had been accustomed to a pioneering lifestyle with little to no regulation on the use of what were formerly open public lands. They especially railed against getting permits from the NPS to do things they had done for decades without oversight. In the newly established Wrangell-St. Elias National Park and Preserve (where I would arrive in the mid-nineties), the hostilities bordered on violent. The first ranger station at Slana mysteriously burned to the ground soon after construction, and the rangers' patrol airplane was set on fire while the rangers were eating dinner in a restaurant in Tazlina. Signs stood outside most restaurants and gas stations in Glennallen that said "NPS not served here," and bumper stickers, with a misunderstanding of biology, announced that "Park Rangers Are Blood-Sucking Maggots." When an African American park ranger stopped in Glennallen to get gas, the attendant shouted, "We don't serve your kind here!" The ranger responded, "Just how far north do I have to go?"

In 1980, the Alaska National Interest Lands Conservation Act (ANILCA) passed Congress and was signed by the president—which somewhat resolved the hunting issues but did not resolve the local opposition or controversies. In one particular incident in 1984, park rangers arrested Joe Vogler, leader of the Alaska Independence Party, for driving his bulldozer across the tundra of Yukon-Charley Rivers National Preserve without obtaining a permit. Vogler was a colorful character, periodic candidate for governor, and an advocate for Alaska to secede from the United States. He disappeared in 1993, just prior to a planned speech before the United Nations where he was expected to call for Alaska's independence from the United States.

Ten years after the bulldozer incident, in October 1994, I was offered the position of Superintendent of Wrangell-St. Elias National Park and Preserve. Before I accepted, I took leave, flew to Anchorage, and drove the 200-plus miles to Glennallen. In a weird coincidence, the one radio station I could hear in my rental vehicle was reporting on the discovery of Joe Vogler's body, wrapped in a blue tarp and buried in a gravel pit near Fairbanks. The radio announcer was interviewing the ranger who had arrested Vogler in 1984, continuing the conspiracy theory that the NPS had been behind his disappearance in 1993. Investigators soon arrested and convicted his neighbor. As I drove along the lonely Glenn highway, I noted that old wounds had not healed and I was heading right into one of the weeping sores of animosity. Some friends suggested I was moving further and further from civilization, but with support from Paula, I accepted the job, fulfilling a career goal of working in Alaska.

Just before Thanksgiving of 1994, my family—Paula and our young two children, Ben and Leah—took the inland ferry up from Seattle and arrived at Skagway in a blinding snowstorm. We drove up the Alaska Highway with the daytime temperature of 30 degrees below zero and about three hours of daylight. We spent our second night in the small town of Tok, population 1,200, and the proprietor of the hotel apologized that she was storing her stuffed Alaska Brown bear in our room. She hoped we would not mind. At

Jonathan Jarvis in Wrangell-St. Elias National Park and Preserve. Credit: Jonathan Jarvis.

first the kids enjoyed taking many pictures with the towering bear in the corner of our room but soon became creeped out. I covered it with a blanket but we still had a somewhat sleepless night. The next morning, as I folded up the electric cords I used to warm the engine block of our truck, the insulation on the wires cracked and fell off into the snow, leaving nothing but a bare copper wire. The owner rightly called me a "cheechako," a Chinook word for newcomer and in Alaska the equivalent to "greenhorn." We stopped for coffee before leaving Tok, which borders the park, and I was immediately recognized as a stranger. There are not many nonlocals in Tok in the winter, so as I was getting my coffee, the waitress quizzed me about why I would be traveling this road, this time of year. When I responded that I was the new park superintendent, a hush fell over the room full of locals. Although the overt threats had subsided over the last decade, there was still an underlying distrust and dislike of the NPS in Alaska.

Wrangell-St. Elias National Park and Preserve is 13 million acres, roughly the size of West Virginia or, for an international comparison, Switzerland. When combined with the adjacent Tatshenshini-Alsek Provincial Park and Kluane National Park in Canada and Glacier Bay National Park, the total 24 million acres makes it part of the largest terrestrial World Heritage site on the planet. Among its "outstanding universal values" are the largest non-polar icefield in the world and high biodiversity in plant and animal communities. The population scattered among small towns, native villages, and lone cabins is around 3,000. It is composed mostly of Native Alaskans, pipeline workers, and families who just want to live a simpler lifestyle. The town of Glennallen, population around 400, has one of the two high schools in the area, one gas station, a couple of restaurants, one hotel, a grocery/general store, a lumberyard/hardware store and a few shops that sell local crafts. It is 200 miles from there to Anchorage by road. It is also considered by many as the coldest place in Alaska, often dropping to 40 to 50 degrees below zero and staying that way for a month or more. An oft-repeated line is that there is "no bad weather in Alaska, only bad equipment," though with all the winter hats, there is a lot of bad hair.

We spent the first few months house-sitting or living in a hotel room because, according to the one real estate agent in the area, "all habitable structures were inhabited." In the spring, we finally bought a house that was great by Alaska standards: it was insulated, had three sources of heat (since at least one will fail), and the water well was under the house so it did not freeze in winter. It also had indoor plumbing, which was a step up from many of the other homes in the area that still required a cold walk to the outhouse! It was walking distance to the school bus stop. And, surrounded by a forest of alder and spruce, it was a short walk to the amazing, wild Copper River.

The only downside was that the water out of our well was not drinkable, due to the local minerals, though it worked fine for showers and toilets. For drinking and cooking, we hauled water once every few weeks from an outdoor community well, which was always an adventure at 40 below zero when any spillage froze to your clothes instantly. Our two children started elementary school in Glennallen, which had a K-12 enrollment of about 300. Paula served as a substitute teacher, organized community service work with high school students, and helped run the park book store. About once a month, we drove the 200 miles to Anchorage to stock up on supplies and enjoy a night at a hotel with an indoor pool. We lived in the Copper Valley for five years, and it was an adventure: moose in our yard, skijoring with our dog, progressive dinners with park families, smoking salmon from the Copper River, challenging the school district to be better than it was, and just enjoying the rural, bush lifestyle. It was hard at times, but both kids made friends and look back on it with warm memories.

Being the park superintendent was more complicated. In this small community, there was no anonymity for me, so I just accepted the fact that I represented the National Park Service, even when I was off duty. I did not hide and conducted my personal business locally as much as possible. My previous time as a park superintendent in rural Idaho at Craters of the Moon National Monument had helped crystallize my appreciation of a rural community. I also know that growing up with my brother in rural Virginia in a traditional hunting and fishing community had set in place an underlying respect for people who lived closer to nature. I knew that many have deep local knowledge and a love of the place. This was common ground where the community and I could come together. I also encouraged my NPS staff of about twenty-five to get engaged in the community: run for school board, volunteer for the fire department, coach sports, or even fix the school computers when they went down. I wanted the community to see us as people, and as contributors, rather than just nameless feds. When I had the opportunity to receive funding for the first visitor center for the park, I made sure it met the needs of the community as well as the park, that the architecture and style met the local vernacular, and that the community would always feel welcome to sit by the fireplace and talk with the staff.

Living in a bush community did give me insights into a wide range of worldviews. The bush often attracts people who "just want to get away," and I learned it was always a mistake to make assumptions about anyone there. The guy in the blood-stained Carhartt coveralls and the sourdough beard could easily have a doctorate from MIT. In the Copper River Valley, the first level of government was the state legislature and the only locally serving elected officials were the members of the school board. As a result, school board meetings were quite entertaining, as all forms of grievances would be aired well into the wee hours of the morning. In a particularly memorable meeting, I gained a sudden insight into the mind of a far-right conservative who had been recently elected to the board. After listening to near unanimous public testimony (I don't remember the topic, only that the audience was unified in its opposition) for quite a while, the board member unleashed a lecture on the group. He said he did not care what we wanted or said, he was "elected to do what *he* wanted, not what the public wanted," and if we wanted to do something different, we should run for office. After the meeting, I went up to him and thanked him for being so honestly undemocratic, as I had suspected this but never really heard anyone say it out loud. I then told him that my wife and I were going to work really hard to get someone other than him elected to the school board.

It became pretty clear to me right away that "managing" a park that is 13 million acres, with just a handful of visitors (about 30,000 per year), was mostly about understanding the resources and working with the local community. The NPS staff scientists were studying the migration of caribou, trying to understand the park's complex geology, and collecting archeological and historical information about human occupation. But for the most part, nature was "running wild" (what former Governor of Alaska Wally Hickel famously warned against).[6] Letting nature do what it does was consistent with the NPS mission and a particularly relevant policy in Alaska. More than once, we pushed back on the State of Alaska Fish and Game, which preferred to manipulate nature by indiscriminately killing bears and wolves to "make" more caribou and moose for sport hunters. This conflict of policy between the state and the feds continues today.

In addition, the Alaska delegation, consisting of Republican Senators Ted Stevens and Frank Murkowski (father of current Senator Lisa Murkowski) and Congressman Don Young, was hostile to the existence of the NPS in Alaska. Stevens, in particular, had led the fight against ANILCA, lost, and was still pissed. That said, as chair of Appropriations in the Senate, Stevens was positioned to help the Alaska parks with infrastructure. He would often beat the crap out of us in a hearing and then, when back in his office, let us know he was sending millions of dollars our way. For instance, I worked closely with Senator Stevens to fund the acquisition of the abandoned Kennecott Mine deep in the heart of the park. In what is known as a split estate, the mine's surface lands (independent of the mineral rights) had been purchased by a group of investors from Anchorage, who were selling off cabin lots on the former mine site, slowly adding to the small population of the McCarthy/Kennecott area. Worried that each of these new park residents would bring with them all the rights of hunting, subsistence, and access granted under ANILCA, I wanted to purchase as much of this private land as possible and make it part of the park. The Conservation Fund worked the deal and Stevens earmarked the Land and Water Conservation Fund for the purchase. I was also able to convince the Kennecott Mine Corporation to donate the underlying mineral estate to the NPS.

Senator Murkowski, who, in my opinion, would rather fight than win, held field hearings just to beat up the National Park Service. In one such hearing, held in Fairbanks with a handpicked crowd representing mostly miners, Murkowski tried to pin me down on denying access to the mining claims in Wrangell-St. Elias, which was untrue. I had a reasonable answer for every one of his questions, and at one point I overheard his staff say to the senator, "stop asking him questions." Congressman Don Young was a rough-and-tumble politician, known for waving a walrus oosik at his perceived enemies, especially those of the Department of the Interior. What I mostly found with these three was that their staff were more hostile than the members of Congress, so I frequently had to put up with being berated by a staffer only to be warmly greeted by the senator or congressman, who seemed to appreciate my relationships with the bush community.

Living in the bush, I had to walk a fine line. The Republican Alaska delegation demanded the NPS should leave or not enforce even the basic rules of a national park. The conservation community demanded that we implement ANILCA as they had hoped it would have been written and as they interpreted it. Caught in the middle was the local community, of which I considered my-

self a member and contributor. I also noted that not a single traditional conservation advocate lived in the bush like I did. While they often chastised me,
or encouraged me to take a stronger stance against the delegation or community, they were never out there to back me up.

When my brother and the other conservation advocates worked the politics to achieve the designation of national monuments under President Carter and, later, parks and preserves under the ANILCA statute of Congress,
they had an opportunity. Smartly, they went for the largest land mass they
could get, drawing boundaries around watersheds and mountain ranges with
the intent to pick up entire ecosystems. This was different from the establishment of national parks in the lower forty-eight, where boundaries often
ran along section lines, state lines, or up against active timber harvest on
the multiple-use lands of the US Forest Service. The parks outside of Alaska
rarely encompassed an entire watershed. This time the concerted effort of
the conservation community and the NPS planners was to get it right from
the start. To begin to act and think at the ecosystem scale was a transforma-
tional act for the National Park Service. In the lower forty-eight (or "outside"
as Alaskans call it) there was deep resistance to such thinking; the NPS was
expected to stay inside park borders.

For Wrangell-St. Elias National Park and Preserve, at 13 million acres the
largest park in the entire system, I came to realize that even this was not big
enough. Salmon spawned in the park rivers but spent most of their lives in
the Pacific Ocean and, as they returned, were subject to commercial harvest
before they reached the park boundary. Caribou herds migrated up the park
mountain slopes to give birth, but in winter they headed out of the park and
traveled into Canada. In the spring, migratory birds came north to nest, rear
their young on the open tundra, and then flew south for the winter. And local
residents of the Alaskan bush also moved in and out of the park for subsistence, firewood, fur trapping, and fishing, by airplane, snow machine, and
all-terrain vehicles. The blinding flash of the obvious was that boundaries
of parks do not encompass entire ecosystems, no matter how big, and that
the porous borders are just lines on a map. As a consequence, I still found a
need to work outside the park as well as inside. This was a revelation for me
and many of us who worked in Alaska.

When President Carter signed ANILCA, the act intended to recognize
the long history of Native Alaskans reliant on fish, wildlife, and other natural resources for their subsistence. From my perspective, recognizing the
thousands of years of stewardship of natural resources by indigenous people
and their traditional ecological knowledge was the right thing to do. In my
previous assignments, I had been well schooled in the history and tragedy of
our nation's treatment of the indigenous people of the continental US, where

native people had died from diseases brought by the pioneers or had been forcibly removed from what became park land. In response, I had worked to reconnect local tribes to places such as Craters of the Moon or North Cascades. Alaska was different, in that the United States never had a war with the Alaska Natives and, for the most part, had respected their traditions, culture, and rights. Some of my most memorable moments there were with native elders. For instance, I was standing outside of the lodge at Dot Lake with a small group of Upper Tanana Athabascan elders. We were waiting for lunch, which was going to be boiled beaver, between meetings on subsistence. One of the men casually mentioned that his grandmother remembered the first time she saw a white man and that he was black. Another time I went to meet the new chair of the Ahtna Corporation. He arrived in a perfect three-piece suit and asked, "Who is your thousand-year planner?" I said, I guess that would be me.

The intent to finally recognize this relationship in statute was partially captured in Title 8 of ANILCA, "Subsistence Management and Use," which describes how the agencies, including the National Park Service, were to allow the taking of fish and wildlife for subsistence within both the national parks and preserves. (The preserves were also open to sport hunting, while the national parks were not; they were known locally as the "park" and the "hard park" respectively.) The original intent of the conservation community was that this subsistence activity would be for Alaska Natives, however a last-minute compromise changed the wording "native" to "local rural." This meant that anyone living close to a national park in Alaska would have subsistence rights to take fish or wildlife within the boundaries of the park.

Although a lot of local rural people are Native Alaskans, there is a significant number of nonnative Alaskans living in proximity to the parks. In the case of the Wrangells, the Alaskan pipeline workers lived in and around small settlements like Glennallen, Gakona, and Gulkana, and thereby could also hunt for subsistence within the national park. There was even a community of fundamentalist Christians who came to Alaska to pursue their faith and preach to the locals, and, as a benefit, could also hunt in the park.

If this was not complicated enough, the intent behind ANILCA was that the State of Alaska would be in charge of implementing subsistence regulations in the parks and public lands, but we soon found that the federal law was in conflict with the state constitution, which allowed for all Alaskan citizens to have equal access to wildlife resources. ANILCA created a different qualification, since Alaskan residents of Anchorage were neither local nor rural and thereby had no subsistence rights.

The conflict between the priority for subsistence by one portion of the

Alaskan population, in particular the native population, and the desires of another portion to hunt and fish as a lower priority, played out in the courts for nearly thirty years. At the center of this conflict was Katie John, an Ahtna elder who took her family annually to a fish camp on a tributary of the Copper River inside the newly designated Wrangell-St. Elias National Park and Preserve. When the state tried to limit Katie's meager harvest of salmon for subsistence, she sued all the way to the Supreme Court and won. This triggered a "federal takeover" of subsistence from the state, over its strenuous objections.

With the federal takeover of subsistence, what was intended by the state was not possible and what was intended by the conservation community was also not possible. The NPS rangers in the field were left to figure it out on the ground. I also had to explain it to the locals! And, as I told many of my staff in my future assignments, unless you have stood in front of a public meeting in an Alaska bar (often the only place large enough) where most of the crowd is drunk and many are armed and yet you deliver a message they really do not want to hear, you just have not done a "real" public meeting. I have to admit that I was never really afraid and actually enjoyed these rather raucous gatherings. I conducted enough public meetings over my five years in Alaska that at one, a guy said that, a few years before, he would have kicked my ass but now he was going to buy me a beer.

What I was not going to do was implement ANILCA as the conservation community intended it to be written, but rather as it was written into law. That meant that local rural residents and especially the Native Alaskans had rights to hunt in the park, and I was going to support that in as fair a process as possible and with fidelity to the statute. Compounding this problem was the often-confusing language of ANILCA, since there was no "conference report" from the legislative debate between the House and the Senate. One of my colleagues in the Alaska regional office suggested that we were all like Talmudic scholars, attempting to interpret ANILCA as Jewish scholars would the Talmud.

After five years in the bush—and with the kids considering whether once they started high school they would want to stay until both finished in the less-than-desirable Glennallen system—a unanimous family vote brought us back to the lower forty-eight. But for those of us who have worked and lived in Alaska, we are never really the same afterward. My appreciation of people who live close to the land deepened and my understanding of the fragility of the Arctic was refined. My experience of assisting native people to preserve their culture as well as working within a large ecosystem shaped my vision in all future positions, including as NPS director. I was the only director of the NPS, to date, to have previously worked in Alaska.

Our Experiences in Alaska Confirm That These
Parklands Are for All Americans

The establishment of national parks, national preserves, wilderness and con-servation areas, and wildlife refuges in Alaska with the passage of the Alaska National Interest Lands Conservation Act by Congress in 1980 is a milestone in the conservation movement not only for the United States but for the world. Like desert lands, Alaska had been viewed as a place to be either exploited or ignored. Changing the minds of entrenched politicians and building public support within the marble halls of Washington could only have been accom-plished by a coalition of conservation organizations representing millions and leadership from the president and key congressional members. It was a rare moment and one that has not been repeated since. As the NPS Alaska Team Captain Roger Contor once quipped, "One hundred people single-handedly saved Alaska."

Today, most Americans, even if they never get there, appreciate the pro-tection of park lands in Alaska and consider it a "win" for conservation. But there are those who see it as a loss to the state for economic development and haven't forgotten. As a result, the presence and management of Alaska lands by the federal government is still a burr under the saddle of the State of Alaska and a regular platform for conservative politics on Capitol Hill. That debate sometimes borders on the absurd, as when longtime Alaska Congressman Don Young famously threatened to whack the Secretary of the Interior's Alaska representative, Deborah Williams, with the penis bone of a walrus. The fact that the representative was a woman was not lost on the congressman or observers.

Those who come into power over the Department of the Interior and the National Park Service who look at the Alaska Lands Act as a loss are inclined to let the Alaskan state government do what it wants over the objections of the NPS (and in many cases, over the laws themselves). This whipsaws the NPS every four years. The state allows hunters to shoot mother bears in the den with their cubs during the winter, arguing that this is a traditional prac-tice that stems from the ancient days of Native Alaskans spearing a bear in a den during the spring "starvation time." While that practice is traditional, it is a far cry from what modern hunters do: enter the den with a high-powered rifle and a bright flashlight. This unsportsmanlike practice in fact is a means to increase the caribou herd, improving the "take" for sport hunters. It con-flicts directly with the stewardship mandates of the National Park Service, and as director Jonathan shut it down. The Trump administration reversed that ruling over the objections of the National Park Service and many conser-vation groups. Indeed, the Trump administration has been one of the most aggressive in recent times for opening up Alaska to development, such as drilling for oil in the Arctic National Wildlife Refuge, logging in the pristine Tongass National Forest, and initially supporting the proposed Pebble Mine and offshore drilling in the Beaufort Sea and Arctic Ocean.

The two of us were honored and privileged to fight for Alaska conservation for much of our careers. Frankly, the work is never done. There will always be political pressure to undermine the conservation accomplishments of each generation. It is our belief that each succeeding generation must then take up the mantle of responsibility and continue the fight. In addition, for the parks in Alaska to truly be preserved for future generations, the National Park Service must be freed from the Department of the Interior, where the forces of exploitation take over every four to eight years. Decades of accomplishments are reversed, such as opening the Ambler Mine Road through Gates of the Arctic or allowing so-called hunters into parkland to kill bear cubs in their winter dens.

Perhaps John Kauffmann said it best in the epilogue of his book, *Alaska's Brooks Range: The Ultimate Mountains*, when he wrote:

> We Americans often regard landmark acts of Congress as great deeds completed, grand dedications following which we need worry no more. Our temperament does not enjoy constant oversight and tending. It is more exciting for us to design than to maintain. Yet tend we must to keep the accomplishments we have achieved. On the ground in Alaska (as elsewhere), we must make sure that the managers of our national lands do their duty, carrying out the plans and enforcing the rules and standards. Virtue does not always come with a green uniform. We should celebrate the NPS professionalism, however, and defend its ranks from political pressure.[7]

THREE

THE POLITICS OF PARK POLICY

Policy without budget is just conversation; show me your budget, and
I will tell you your policy.

GEORGE B. HARTZOG, JR., NPS Director 1964–72

There is likely no drier topic for a chapter than management policies, but for
the National Park Service, *Management Policies* is an official document: Ikea
assembly instructions, cookbook, and "How to Run a Park for Dummies"
wrapped into one volume. On the desk of every park superintendent, within
easy reach, it is dog-eared, coffee-stained, and tabbed to the frequently used
pages. When confronted with a new challenge or issue, it is the go-to manual to start the decision process. Sometimes the answer is there, and, if not,
then it provides a framework to make a decision. When the NPS *Management Policies* is applied consistently across all the national parks, then there
is a system that is far greater than just the sum of the parts.

Too often in the history of America's national parks since 1972, policy from
one administration to the next has shifted widely, either through ignorance or
malice, bad judgment or intentional distortion, or just plain politics. National
Park laws are rewritten, or attempted to be rewritten, by Congress and are
reinterpreted by administrations. Then regulations must be rewritten. New
policies are issued. Often lawsuits follow, and courts render decisions determining how an NPS law is to be interpreted and enforced. Through its more
than one-hundred-year history, the NPS has sought to develop and maintain

a clear set of policies that apply to the whole system. Today, the *Management Policies* manual provides that cohesive framework.

Over our years working with and within the NPS, we have personally participated in reviewing, commenting on, and rewriting sections of successive editions of *Management Policies*: in 1978, 1988, 2000, and 2006, which was the last time it was officially rewritten. Too often, the proposed, politically motivated swings in policy have been breathtaking, most notably in the Reagan and Trump administrations, as we will describe. Fortunately, the most radical proposals to shift policy away from the Organic Act have been thwarted by a combination of internal and professional resistance combined with external outright opposition, at times with vigorous support from key members of Congress.

After law and regulation comes policy, which is the agency's interpretation of regulation and law. Policy is written in terms that can be understood by field personnel and often gives some flexibility in implementation. Shifting winds of politics applied to policy have too often led to erosion of the professional management of the parks. Management policies, however, are not regulations, and are not enforceable through litigation; they only apply to the career employees of the NPS and then only to the extent that senior leadership requires. Nevertheless, the NPS *Management Policies* expresses the agency's intent. It is intended to be the bible of the NPS.

The first NPS director, Stephen Mather, had the unique opportunity to work with Congress to develop the specific language of the NPS Organic Act, with its critical statement of purpose for the parks, "to conserve unimpaired . . . for enjoyment of future generations." He then drafted the first set of NPS policies that interpreted the Organic Act. Mather wisely gave this policy draft to his boss, Secretary of the Interior Franklin Lane, who sent it back to Mather in a formal policy letter, still referred to today as the Lane Letter (March 1918).

After the Lane Letter, policy evolved slowly, changing as the types of places added to the system changed. DOI Secretary Hubert Work essentially reissued the Lane Letter in 1925. However, due to rising conflict between the NPS and the US Forest Service, stemming from the fact that lands for new national parks were being transferred from the USFS to the NPS by acts of Congress, Work added a section explaining the difference in purpose between national forests and national parks, directing the NPS to confer with the USFS before recommending any USFS lands for addition to the national park system. This policy had a (temporary) chilling effect on the carving out of new national parks from national forests. The intense conflict between the NPS and the USFS continues to the present day.

The first set of formally adopted NPS guidelines, a three-page, seventeen-point document entitled "General Policies," was developed by DOI attorney (and former Congressman) Louis C. Cramton and incorporated into Director Horace Albright's annual report to the Secretary in June 1932. It was a slightly more detailed expansion of the Lane Letter, reinforcing the inherent mission challenge by referring to the "twin purposes" of the national parks as "enjoyment and use by the present generation, with its preservation unspoiled for the future." The public use of the parks referred to by this policy, however, noted that *"education* is a major phase of the enjoyment and benefit to be derived by the people from these parks" and that "recreation, in its broadest sense, includes much of education and inspiration. Even in its narrower sense, having a good time, is a proper *incidental* use" (emphasis added).

The role of visitor education in the national parks as an essential element of the NPS mission was not established in law, rather than merely policy, until 2016. However, the role of active recreation, and its "balance" with preservation of park resources, has been debated in every administration and under every director and Secretary since Mather and Lane.

In 1940, in an official "Brief History of the National Park Service," the NPS noted that "proper administration of the national parks will retain these areas in their natural condition, sparing them the vandalism of improvement." This recognition was not always easily supported, even at the time. Newton Drury's tenure as NPS director, 1940–1951, spanned the years of World War II when parks were threatened with exploitation to support the war. His policy admonition to the country was that "if we are going to succeed in preserving the greatness of the national parks, they must be held inviolate. They represent the last stands of primitive America. If we are going to whittle away at them, we should recognize, at the very beginning, that such whittlings are cumulative and the end result will be mediocrity."

When wilderness areas were overlaid on portions of existing undeveloped lands in some parks by acts of Congress beginning in 1964, this represented the first (but not the last) time that Congress sought to order specific management policy direction for the national park system by permanently "zoning" these areas to prohibit development.

Political controversy began to creep into management policies shortly after the July 1964 "Memorandum to the Director of the National Park Service from the Secretary of the Interior on Management of the National Park System," an initiative of Director Hartzog, which, for the first time in official policy, directed the NPS to divide the system into three categories for management—natural, historical, or recreational. The statement from the Secretary further stated that recreation should be the dominant or primary objective for sites in the recreation category. The NPCA objected to the lumping of all such sites into a recreation category in an editorial in the September 1964 issue of *National Parks Magazine.*

At the same time, the NPS and citizen advocates across the country, led by the NPCA and others, were considering possible new units of the system along coastal seashores and the shores of the Great Lakes, including Fire Island, Assateague, Cape Lookout, Cumberland Island, Canaveral, Gulf Islands, and Padre Islands along the Atlantic and Gulf; Point Reyes on the Pacific coast; and Pictured Rocks, Indiana Dunes, Apostle Islands, and Sleeping Bear Dunes along the shores of the Great Lakes. All of these sites were added to the system between 1964 and 1972 under Director Hartzog's active leadership. None of the citizens advocating for protecting these ocean and lake shores wanted to see mass recreation developments there. The public wanted something other than another Coney Island. Accepting that these new parks were not all pristine landscapes, and that active recreation use was a purpose for their designations, park advocates nevertheless urged that recreation must be fully compatible with protecting the resource values of these places.

On June 18, 1969, the NPS published its first comprehensive, detailed set of policies, entitled *Policy Guidelines of the National Park Service,* which was written by the agency, published by the Secretary, and delivered back to Director Hartzog. But by early 1970, the NPS confronted an increasingly diverse set of places to manage, especially places emphasizing outdoor recreation. Director Hartzog, responding to direction from the Secretary, divided the system into three distinct sets of management policies. According to these new "management categories," the system was partitioned into natural areas, historic areas, and recreation areas, with a corresponding policy book for each.

DESTRY

The advent of these management categories, especially lumping all of the rivers, trails, seashores, and lakeshores into the recreation category, sparked a great deal of controversy from conservation constituencies. This issue was the first big conflict that the NPCA had with the NPS after I joined the staff in 1972. The NPCA and other park constituents had lobbied Congress especially hard for establishment of these seashores and lakeshores as units of the national park system. We did not want to see them dominated by recreating hordes and new recreation developments detrimental to wildlife or that would diminish the visitor education, appreciation, and inspiration that national parks have traditionally provided.

Joe Zysman at Fire Island National Seashore, Otis Pike Wilderness. Credit: Destry Jarvis.

By early 1973, President Nixon had fired Hartzog, and, with four NPS directors appointed over the next seven years, the revolving door at NPS headquarters hardly allowed time for much change in management policies. Nevertheless, during these years, I led an effort to eliminate the divisive management categories among various national park advocates who saw them as inconsistent with the intent of Congress in adding diverse sites to the national park system.

A perfect example of the conservation concerns with putting all of the seashores into the recreation category emerged in 1974. The NPS produced a new master plan for Fire Island National Seashore in New York, without any public involvement or comments solicited, that proposed building new "beach complexes" every 1.5 miles along the eight-mile so-called natural zone of the seashore. With intense recreation development already in place in state parks at each end of the seashore, I and others, including the Fire Island Wilderness Committee led by Joe Zysman, strongly opposed this new development plan. Instead we proposed the eight-mile natural area be designated by Congress as wilderness—to protect it from the prodevelopment forces within the National Park Service. Joe and I met with the DOI assistant secretary for Fish and Wildlife and Parks, Nathaniel P. Reed, to point out the numerous objections to the NPS development plan. Nat Reed, a Republican and lifelong conservationist from Florida, was one of the top two or three assistant secretaries ever to supervise the NPS in its history, and his numerous park-related achievements will endure in perpetuity. Sadly, Reed died in July 2018 while fly fishing, one of his many passions.

After much additional public protest against the NPS plan, Reed ordered the NPS to throw the Fire Island Plan into the trash and start the planning process over, with full citizen engagement this time. While Joe and I worked in Congress for the next six years to have the eight-mile natural zone designated as statutory wilderness in 1980, this part of the seashore was given interim protection by the NPS as a result of citizen action and with Reed's full support.

When a similarly egregious new master plan was produced by the NPS for Yosemite National Park, including a concessionaire-advocated tram ride for visitors from Glacier Point into the valley, Reed ordered the NPS to completely revise its whole planning and compliance process. He put John Reynolds, a rising NPS leader, in charge of the project. John revolutionized NPS planning, especially by requiring extensive public involvement early in the scoping phase. He went on to be superintendent of North Cascades National Park, director of the service-wide planning and design center in Denver, Northeast regional director, and deputy director of the service during the Clinton administration.

In 1975, the NPS was engaged in a back-and-forth dialogue with the House Interior Committee, especially Representative Phillip Ruppe (R-MI), regarding the application of the recreation category to various national seashores and lakeshores. NPS Director Gary Everhardt defended the categories while trying to back away from strict adherence to them. In his June 4, 1975, letter to Assistant Secretary Reed, Representative Ruppe noted that "the language of the laws authorizing such areas at Canaveral National Seashore, Big Cypress National Preserve, Big Thicket National Preserve, and Cape Lookout National Seashore seems implicitly clear that these areas if categorized, should fall into the natural category. Yet the Code [of Federal Regulations; CFR] definitions would not permit this in the case of the seashores." In reply to Ruppe's letter, Everhardt wrote, among other things, that "we intend to amend out of existence statements such as those in the CFR 1.2.(1) which state that parks in the recreation category are 'administered primarily for the purpose of public recreation.'" To his credit, Everhardt did issue a "new" policy statement essentially restating the interpretation of the 1916 Organic Act's NPS mission, that "in the planning and management of all parks we must be guided by the unifying principle that protection of ecological health and historic integrity is our first consideration. . . . Thus park uses must be limited to those activities that are dependent upon and protective of the natural and historic values each park was established to preserve." But he did not succeed in "amending out of existence" the conflicting language before he was replaced as director early in the Carter administration.

In 1976 and early 1977, I prepared, on behalf of NPCA and other national conservation organizations, an extensive analysis of the management plans for each of the national seashores and lakeshores in order to point out their inadequate resource protection under the recreation management category. This analysis covered Cape Hatteras, Cape Cod, Point Reyes, Padre Island, Fire Island, Assateague Island, Cape Lookout, Gulf Islands, Cumberland Island, Canaveral, Pictured Rocks, Indiana Dunes, Apostle Islands, and Sleeping Bear Dunes.

In 1977, newly installed NPS Director Bill Whalen, previously the superintendent of one of the urban national recreation areas, Golden Gate NRA, took up this issue in earnest. In early October, I prepared a "Petition for Rulemaking" seeking repeal of the regulation that established the recreation category, and I shared a draft with Director Whalen. The very next day, he sent me a letter seeking NPCA opinion on a draft memo he had prepared for Assistant Secretary of Fish and Wildlife and Parks Robert Herbst, seeking approval to repeal the recreation category. Whalen's letter to me stated:

> I would appreciate your views regarding the enclosed memorandum, which I propose to send to Assistant Secretary Herbst for his concurrence. With his approval, I intend to issue a directive to all NPS managers to the effect that the use of management categories has been rescinded. We will take steps to eliminate reference to categories in all NPS policies and other documents.[1]

I heartily endorsed Director Whalen's proposed policy change. Throughout Whalen's brief tenure as director, I met with him on a regular basis to discuss management policies as well as the draft that became the 1978 edition of the *Management Policies* manual. To his great credit, Whalen also organized and regularly convened meetings with several conservation NGOs that came to be known as the Unity Group, which I helped schedule and coordinate.

Assistant Secretary Herbst approved the change to eliminate the arbitrary separation of units of the system into categories. All units of the system would be managed under a common set of policies. However, whatever "policy" one administration adopts can be changed by the next, as we have seen many times since then. Thus, new legislation was needed in order to make this change in policy the new statutory law.

In 1978, I worked closely with House Interior Committee staff on amendment language to the 1970 NPS General Authorities Act which Congress then enacted, largely in response to public policy disagreements with the management categories policy. These amendments "put a thumb on the scales" of NPS policies, pushing the service further toward the predominance of resource preservation within a unified system. The original language of the General Authorities Act stressed congressional intent with regard to the national park system having integrity as a whole greater than the sum of its units:

> Congress declares that the national park system, which began with establishment of Yellowstone National Park in 1872, has since grown to include superlative natural, historic, and recreation areas in every major region of the United States, its territories and island possessions; that these areas, though distinct in character, are united through their inter-related purposes and resources into one national park system as cumulative expressions of a single national heritage; that individually and collectively, these areas derive increased national dignity and recognition in their superb environmental quality through their inclusion jointly with each other in one national park system preserved and managed for the benefit and inspiration of all the people of the United States.[2]

On March 27, 1978, Congress added another amendment to the General Authorities Act of 1970. (This particular amendment was inserted into a new statute expanding Redwood National Park, on which I worked on with Congressman Phillip Burton; see chap. 1). This provision, which has come to be referred to as the nonderogation standard, stated that:

> The authorization of activities shall be construed and the protection, management, and administration of these areas shall be conducted in light of the high public value and integrity of the National Park System and shall not be exercised in derogation of the values and purposes for which these various areas have been established, except as may have been or shall be directly and specifically provided by Congress.[3]

Basing their interpretation on the original language of the statute and the 1978 amendment, NPS Director Bill Whalen and his policy staff, aided by DOI attorneys, determined that the 1916 Organic Act applied to every unit of the system, the application of which to any specific unit only differed if that park's individual enabling statute specified otherwise. When the expanded *Management Policies* manual was released later in 1978, it was some 150 pages, up from a maximum of ten pages in previous versions.

Although the NPS's own professional understanding of its mission, as reflected in *Management Policies*, has changed very little through its various editions, the Reagan and both Bush administrations attempted to fundamentally alter the NPS interpretation of its mission in the 1988 and 2006 editions.

These battles, waged by politically appointed assistant secretaries for Fish, Wildlife and Parks and their political staffs, were highly charged, fraught with conflict internally and externally, and damaging to the NPS.

Soon after President Reagan nominated Secretary of the Interior James Watt in spring 1981, Watt gave an infamous speech to the Conference of National Park Service Concessioners in which he said a number of startling things, the most relevant of which for NPS *Management Policies* was "I will *err on the side of public use versus preservation.*" Up until that time in NPS history, no Secretary had been so bold as to contradict the clear intent of the NPS 1916 Organic Act. The Secretary went on to say, "We will use the budget system to be the excuse to make major policy decisions."[4]

Secretary Watt retained Russ Dickenson as NPS director. A career NPS professional, Dickenson had served as NPS National Capital regional director during the Nixon administration (when Watt was director of the Bureau of Outdoor Recreation) before becoming NPS director in the last two years of the Carter administration. Though nothing was immediately done to revise NPS policy as a result of Watt's speech, the Secretary continued to generate widespread controversy with his policies and extreme statements. Finally even President Reagan had heard enough and fired him in early 1983, replacing him with a more benign lawyer, William Clark. NPS policy had sustained no major changes except for various budget and staffing reductions.

After Reagan's second term began in 1985, Deputy Secretary Don Hodel was promoted to be Secretary of the Interior; Assistant Solicitor William Horn was confirmed as assistant secretary for Fish and Wildlife and Parks; and William Penn Mott was chosen as NPS director. Bill Mott had been California's State Parks director when Reagan was governor, and he was a member of the NPCA's board of directors when he was chosen. Mott and Horn clashed strongly from the outset, especially over interpreting the core management policies of the NPS. When Bill Mott first flew to DC from his West Coast home, I picked him up at Dulles Airport. On the drive into the city, Bill pulled an envelope from his pocket on which he had written the draft of his "Ten-Point Plan" for the NPS improvements he would seek. His plan was totally consistent with long-standing NPS interpretation of the meaning of the Organic Act and the preservation mandate.

Very soon after assuming their new positions, Horn wrote a memo to Mott launching his nearly four-year effort to fundamentally alter the NPS interpretation of the 1916 Organic Act, which had stood for nearly seventy-five years at that point. Horn's December 1985 memo asserted that a "fundamental consideration" for park natural resources management was that "natural features are conserved chiefly for the benefit, use and enjoyment of the general public. . . . *Given that public recreational benefit is the principal rea-*

son for conserving natural features we must be very judicious in considering a feature to be of national significance solely on the basis of its import to the scientific community" (emphasis added).[5]

Horn's memo went on to urge the NPS to distinguish among each individual park's natural features as to which were of international, national, regional, or local significance. "We must also be prudent in deciding what is and what is not essential to achieving the main mission of an individual unit. For example, the aesthetics of Yosemite Valley are obviously essential to the park whereas the health of the mule deer herd is not."[6]

For me at the NPCA, this gross misinterpretation of the NPS 1916 Organic Act was tantamount to a declaration of war on the national parks as they had been managed for more than a hundred years, so we launched a four-year campaign to stop Horn's efforts. Fortunately, we had the NPS director on our side, and Mott's past close relationship with President Reagan ensured that Horn couldn't fire him, however much he wanted to do so.

After seeking to have NPS redefine "significance" of park resources, Horn's memo sought to have the NPS rate each park natural resource "condition" as good, fair, or poor, noting that "change" does not equate to "damage." It urged the NPS to drop reference to resource "threats," renaming such matters "adverse actions," and noting that an action deemed adverse in one park may be acceptable in another. "For example, the fact that it will introduce an exotic species is not prima facie evidence of damage, for the species itself may not cause damage."[7] Horn's management policy would have reversed NPS longstanding policy that manages the natural area parks on a holistic basis, not focused on individual species but on their interdependence, intact habitat requirements, and natural processes.

On February 18, 1986, Horn sent Mott a formal memo, urging that the new edition of *Management Policies* should be "a much shorter document than the present version" and completed by May 1, 1986. In addition to eliminating any opportunity for public review, his approach would have shifted NPS policy focus to broad flexibility for each individual park and away from service-wide policies that view the system as unified. Horn's memo also stated that "the revision should be guided by the following principles: Resource protection and providing for visitor use and enjoyment are co-equal mandates."[8]

On February 24, 1986, NPS Deputy Director Denis Galvin formally responded to the Horn memo, noting that he had canvassed all of the NPS regional directors and concluded that:

> There was essentially unanimous agreement that the "Management Policies" did not require major or extensive revision or change. The present "Management Policies" document has served the Service very well, and we do not

believe any useful management need would be met by significant change or revision, either as to style or content. . . . Regarding your suggestion that the "Management Policies" be redirected to distinguish between types of National Park System units, we strongly oppose such an approach. . . . Congress made it expressly clear in the Act for Administration (16 USC Section 1c) that all units of the National Park System are equally protected by law without regard to the various titles by which Congress has chosen to name them.[9]

Horn's February memo was leaked to me soon after it was sent, and on March 6, I issued a public statement for the NPCA that defined his proposal as "disastrous new national park policy," noting that

> This memorandum from the Assistant Secretary's office is one of the most outrageous distortions of existing law and park policy that we have ever seen, and represents a fundamental and erroneous misinterpretation of the mandates of the Congress for the administration of the national park system. . . . In NPCA's view, the only possible purpose for this sort of direct distortion of the facts which this memo represents is to afford the Assistant Secretary's office the opportunity to argue that none of the parks are "threatened" as a means of contrasting with NPS's State of the Parks Report to Congress in 1980, which accounted for over 4000 specific threats to parks throughout the system.[10]

On May 15, 1986, Deputy Director Galvin again sent Horn a memo, in response to a meeting that they had had the previous week. He again rejected Horn's concept of managing each national park unit solely based on its individual enabling statute, noting that:

> Titles or other classifications assigned on a park wide basis have no useful relationship to management policy or practice. Though NPS tried to fashion management policies around a three-category classification system in the 1960s, that system proved both controversial and cumbersome. In the end, it was abandoned as unworkable.[11]

On June 6, Horn sent Mott another memo, this time spelling out a schedule for revisions to *Management Policies*, especially requiring a revision to chapter 1 on the fundamental purpose of the parks, by September 22.[12]

On July 15, Director Mott sent Horn a reply, stating "We clearly have some substantial differences in perspectives and points of view regarding the management policies affecting the National Park System. But rather than attempt to resolve those issues by continuing this exchange, we propose to develop a new first chapter for the revised 'Management Policies.'"[13]

After a series of internal staff meetings between the NPS and the assistant secretary's office over the summer, Director Mott sent Horn a memo on September 11, noting the assistant secretary's insistence that the new edition of *Management Policies* "expand management flexibility," specifically noting that it was Horn's intent that the new policies

- be made more flexible for exotic species
- be more liberal for ORV use at national seashores
- be more permissive of permitted uses at national recreation area sites
- improve public access at national recreation areas through road construction
- give more policy flexibility on use of pesticides at units other than national parks[14]

On September 30, Director Mott sent Horn a detailed draft rewrite of chapter 1 of *Management Policies*. This draft rejected Horn's unit-by-unit approach to policy as well as his preference for reinstating several management categories. In particular, the draft continued to rely on the basic statutory interpretation of the 1916 Organic Act, noting that:

> All units of the National Park system are equally significant. There are no lesser and no greater parks—only different parks, serving to protect different resources in different locations. . . . All units of the National Park System are nationally significant. . . . Qualitative distinctions among units of the National Park System are not recognized as having a valid basis in National Park Service policy.[15]

That same day Horn replied to Mott with a memo that stated the draft did not satisfy his requirements and a responsive reply was overdue. He specifically stated that:

> As for categorization of parks, NPS has yet to make a serious effort to develop categories pursuant to this initiative. Perhaps there should be three categories or four or more. . . . I need a complete briefing on at least three different approaches to categorization—including the parks that would be included in each category in each approach as soon as possible but no later than October 31, 1986.[16]

Several more drafts of chapter 1 were prepared by the NPS over the fall, going back and forth between the NPS and the assistant secretary. On November 25, 1986, Horn sent Director Mott yet another memo, this one accompanied by his handwritten margin notes on the draft. Horn pointedly stated:

> The staff doing the drafting should review my earlier memos on this topic so that the policy direction therein is clearly and unambiguously reflected in this chapter; that is not presently the case. . . . In preparing the new draft it is critical to keep in mind that the statutory language in the Organic Act speaks to conserving resources for the enjoyment of the public. Public enjoyment is the objective of resource conservation as envisioned by Congress. It is not, as suggested by the draft chapter, a mere stipulation. The record does not support the notion that Congress ever envisioned the parks to be elitist preserves open only to the self-selected few.[17]

Horn focused in particular on his attempt to redefine "impairment" of park resources, noting that the term is highly subjective:

> It must first and foremost be related to physical resources themselves and must not be invoked to undermine the intent of Congress to provide for public use and enjoyment. Consequently "impairment" should be thought of principally in terms of something that would result in lessening of present and future visitor enjoyment. . . . Key is to not impair the physical resources as opposed to highly subjective amenity values.[18]

I had lunch with Bill Mott on December 3, a day after his latest meeting with Horn. Bill reported that he and Horn were at an impasse, with Horn insisting that he would move forward with the policy changes, and Mott likewise insisting that he was opposed and would not agree. Bill noted that he did not have support from Secretary Hodel, and that the NPS would have to go through a public process, soliciting comments from the public, the conservation community, and Congress, and engage the news media—all ways to thwart Horn's attempt to rewrite NPS policy.

Up to this point, all these discussions had been internal to DOI, with the occasional leak of drafts to me at various times over the summer and fall of 1986. On December 5, *New York Times* reporter Phil Shabecoff,[19] with whom I had been engaged in a regular discussion about the inner politics of the DOI/NPS for two years, called to say he was going with the story, which appeared on December 6. His article not only led to a cascade of similar stories around the country but had the directly beneficial effect of greatly slowing down Horn and his fellow political appointees on their radical revisions to *Management Policies.*

What did not slow down was Horn's insistence on a major shift in NPS senior personnel in retaliation for the internal opposition to his policy changes, also over Mott's strenuous objections. These forced moves of NPS senior professional career managers harkened back to former Secretary Watt's 1981 statement that "if you have a problem with NPS, I will change the policy or the person, whichever is easier." Horn, five years later, was attempting to likewise change both NPS policy and professional personnel.

The NPS returned *Management Policies* draft #7 to the assistant secretary on December 12, 1986, with some of the fundamental changes that Horn had sought. At this time, there still had been no circulation of the draft of *Management Policies* to the field offices of the NPS or to the public for comment.

On March 8, 1987, the NPS circulated an internal memo to senior leadership covering *Management Policies* draft #9, stating that it had been approved by both the director and assistant secretary. Director Mott was still opposed and refused to sign the memo. Instead it was signed by newly installed NPS Associate Director for Policy, Management, and Budget Ed Davis, an obviously politically motivated and entirely new senior position.

I fired off a letter to Director Mott on June 9, supporting his position in opposition to the fundamental changes being proposed by Horn. "We feel that the most recent draft that we have seen, draft #9, is significantly different from both the 1975 and 1978 Manuals. . . . The concept of conservation as an end in itself is virtually ignored in draft #9 of Chapter One, although it is clearly specified as a primary function of the Park Service in both the 1975 and 1978 Manuals."[20]

However, on June 26, 1987, I received a reply from Associate Director Davis, declining the NPCA's comments. He noted, "since we are still in the process of reviewing and revising our management policies, to date none have been released to the public. We anticipate that we will have a draft document ready for review and comment by fall of 1987 . . . at that time we will welcome your comments."[21]

It was not until March 25, 1988, however, that notice appeared in the *Federal Register* (53FR 9821) that proposed revisions to NPS *Management Policies* were available for public comment. I prepared detailed, section-by-section written comments on the draft revisions to *Management Policies*, submitted May 27, 1988.

In general it appears to us that this draft of the Management Policies is an across-the-board watering-down of the philosophy and ideals of the National Park System. . . . In conclusion, we are well aware that the NPS has gone through an excruciatingly long and laborious process in bringing out the proposed draft of Management Policies, and that the professionals of the Service beat back attempts by the Administration appointees for even worse erosions of policy than are contained in the version published. This perseverance by you and your policy staff is to be commended. However, the deletions, distortions, and outright reversals of existing policy and Service traditions contained in the proposed draft mandate a retraction and restart of the entire process.[22]

Given the proximity of the timing of the policy review to the start of the annual summer vacation season, my comments to the media about this draft of management policies were:

As Americans travel to their national parks this Memorial Day weekend, behind the scene forces are at work to fashion this nation's parks into low-quality recreational playgrounds which would jeopardize the world's pre-eminent park system. This so-called revised Management Policies manual, currently under public review, is nothing but a final assault by this Administration to de-emphasize the Park Service's mandate to preserve our Nation's heritage.[23]

More important than my comments for the NPCA, on June 10, 1988, Congressman Bruce Vento, chair of the House National Parks Subcommittee, sent Director Mott a very strongly worded letter in opposition to the proposed changes to *Management Policies*, a letter which I, working with his staff, had helped to craft. He noted that:

Congressional intent has long been quite clear that the preeminent responsibility of the National Park System is the preservation of those natural and cultural resources so recognized. The 1916, 1970, and 1978 Acts all specifically recognize this prime responsibility. . . . These [draft] policies should not be finalized as they are contrary to the legally defined mission of the National Park Service. . . . I do not accept this draft as congruent with the legislative intent establishing the National Park Service. . . . I am particularly concerned by the Chapter One discussion on "impairment." Few words in the English language have less ambiguity. The mission of the National Park Service is to protect the System's resources *unimpaired*.[24]

Similarly, NPS park superintendents sent in comments to the director, also objecting to the proposed revisions. Mammoth Cave Superintendent Dave Mihalic wrote:

The statement that "only in those instances when impairment is a consequence of current activity does the conservation mandate take precedence" is a dangerous and radical departure from the focus, intent and purpose of the agency. . . . This literally puts the agency behind the curve in those areas in which waiting for impairment to manifest itself may be too late.[25]

Everglades National Park Assistant Superintendent Rob Arnberger wrote:

As written, impairment of a resource is defined in terms of overuse or improper use, implying in-park actions. The greatest threats to the parks, now and in the future, will lie in impairment of park resource values by external forces. These external forces will not be over using or improperly using park resources, but instead, will be detrimentally affecting the ecosystem processes within the park.[26]

By the time the final version of the new edition of *Management Policies* was released in December 1988, two-and-a-half years after Horn's initial deadline, all of the most egregious changes that he had pushed Director Mott for had been removed or diluted, so that this edition did not differ philosophically from the previous edition of 1978. The 1988 edition was, however, not as explicit on the application of the impairment standard for resource protection as was needed, leaving much discretion to the individual park superintendent. For example, in chapter 1.3 of the 1988 edition, the NPS states that "There will inevitably be some tension between conservation of resources on the one hand and public enjoyment on the other. The National Park Service is charged with the difficult task of achieving both."

Another difference in the new edition was a clear recognition that individual enabling statutes that authorize the NPS to manage and protect a unit can modify the application of the 1916 Organic Act to that park unit if it specifically so states. The Organic Act still applies to every unit of the system, but Congress can, and sometimes does, allow differences.

Also, by this time, due to intense public pressure and threats from Congress to abolish his position and/or zero out his salary—including from parks subcommittee Chairman Bruce Vento and Interior Appropriations Subcommittee Chairman Sidney Yates, both of whom were strongly opposed to Horn's proposed actions and policy changes—Horn had been forced to resign. George H. W. Bush had been elected president. And during his term in office neither Secretary of the Interior Manuel Luján nor NPS Director Jim Ridenour sought to amend *Management Policies* during their tenure, from 1989 through 1992.

From the outset of the Clinton administration in 1993, NPS leadership expressed its intent to issue a new edition of *Management Policies*. In keeping with recent trends, such a new edition was targeted for issue in 1998 (following the 1978 and 1988 editions), and work began in earnest under the NPS chief of policy, Bernard "Chick" Fagan. However, two major issues intervened and slowed the process.

First, beginning early in 1993, Vice President Al Gore launched his major "Reinventing Government" initiative, which consumed much of the time and energy of NPS Director Roger Kennedy between 1993 and 1996.[27] Second, DOI leadership formulated the idea of promulgating an "impairment rule," a formal federal regulation to be published in the *Federal Register*, as the way to make the Organic Act's standard of managing park resources to remain "unimpaired" enforceable as a matter of law. NPS leaders saw the need to amend *Management Policies* in order to strengthen their own requirement to avoid impairment, and to state in clear, precise terms, that protecting park resources is, under the law, the first priority for NPS managers.

The DOI solicitor's staff attorney, K. C. Becker, and Deputy Assistant Secretary for Fish and Wildlife and Parks Stephen Saunders worked in earnest from 1998 throughout 2000 to craft a formal rule-making that could secure approval of the proposed impairment rule. It would need concurrence from DOI Secretary Bruce Babbitt, the Justice Department, and approval from the White House Office of Information and Regulatory Affairs (OIRA, part of the White Office of Management and Budget).

Throughout much of 1999 and 2000, NPS staff, the assistant secretary's staff (which included myself), and the solicitor's office undertook review and rewriting as needed of the *Management Policies* manual. Initial emphasis was on chapter 1, now called "The Foundation." Our goal was that chapter 1 would accomplish two very important clarifications that were not made in the 1988 edition of *Management Policies*. First, it was necessary to develop a clear and definitive distinction between resource "impact" and resource "impairment." Impacts are permissible in parks, as long as they are not causing impairment. Second, this chapter had to clearly articulate the fact that the proper interpretation of the language of the Organic Act places resource protection from impairment as the primary mission of the NPS; only that visitor enjoyment of the parks that does not cause impairment is permissible.

From the position of the authors, many scholars who study NPS history, and from a close reading of *Management Policies* and court rulings, the NPS does not have a "dual" (as has often been asserted) mission of two equal components, preservation and use. Rather, resource protection and prevention of impairment of resources takes precedence whenever there is conflict between these two elements of park management. As we began to develop various drafts of the needed language for the new edition of *Management Policies*, there was total agreement on the objective but lots of opinions about which words to use.

To illustrate this point, Assistant Secretary Don Barry sent NPS Director Bob Stanton a memo on August 9, 1999, which I helped him to prepare. It concluded, "However, I believe the correct reading of the Organic Act clearly indicates that there is in fact, only one mission. The NPS must allow for visitation in the parks, but must manage it in such a manner as to conserve unimpaired the natural and cultural resources of the parks for future generations."[28]

Similarly, in further rejecting the idea of dividing the system into "management categories," Assistant Secretary Barry's memo stated:

We do not have different sets of policies for units with different resources. While there may be direction in some park-specific legislation that modifies the manner in which a unit is operated under the Organic Act, without such specific direction, each park unit regardless of its designation, is to prevent impairment to both natural and cultural resources.[29]

In addition, Barry's memo asserted that "Finally, I would ask that every qualifying phrase such as 'whenever possible' or 'to the greatest extent practicable' be thoroughly examined, and removed unless it is found be absolutely necessary."[30]

Over the remaining months of 1999, the draft of chapter 1 of *Management Policies* went through a dozen revisions, circulated and commented on by most senior staff in the NPS and by myself and others in the assistant secretary's office. On January 7, 2000, Secretary Babbitt signed off on NPS Director Stanton's policy memorandum entitled "Interpretation of the NPS Organic Act and the NPS General Authorities Act." The memo thus formed the definitive policy to be included in *Management Policies* on how the fundamental laws of the NPS were to be understood and applied by all park managers.

The memorandum concludes that:

This interpretation should end any confusion over whether these laws give the NPS equal responsibilities to provide for the conservation of park resources and values and for their enjoyment. This interpretation makes it clear that, when there is an unavoidable conflict, conserving those resources and values is predominant.[31]

Of equal importance, it was clear that NPS policy must distinguish between resource "impacts" and resource "impairment"—in order for field managers to make decisions, they need to understand and make scientifically informed decisions about which impacts are acceptable or can be mitigated, and which would cause impairment and must be avoided.

Notice was posted in the *Federal Register* in March 2000 soliciting public comments on the proposed text of *Management Policies*, and many comments were received, reviewed, and considered. After that, internal debate continued with many additional drafts of language circulated through the fall of 2000.

By then, it was clear that the White House Office of Information and Regulatory Affairs would not support the NPS in promulgating an impairment rule. Consequently, on November 17, Acting NPS Director Denis Galvin signed Director's Order 55, "Interpreting the National Park Service Organic Act," so that the clarified interpretation of impairment would at least be mandatory on NPS field managers and enforceable on NPS in federal court. This Director's Order was then superseded by the final version of the complete new edition of *Management Policies* in early 2001, before the change in administration.

The impetus to get the new edition of *Management Policies* signed and published was threefold—to improve the weaker 1988 edition, to get the new edition finalized before the change in administration to that of newly elected President George W. Bush, and to clearly set NPS policy in light of pending federal litigation in several cases that invoked interpretation of the Organic Act, including a major case in Utah's Canyonlands National Park.

Canyonlands National Park finalized its general management plan in 1972, eight years after the park was established by Congress. It stipulated that the prepark use of some 200 miles of dirt roads and trail running through the park would remain open to off-road vehicle (ORV) usage. However, as the years passed and ORV use of these trails expanded dramatically, NPS Park Superintendent Walt Dabney recognized the need for a backcountry management plan that specifically addressed the impacts, location, and extent of ORV use in the park.

In December 1993, the NPS released for public comment an Environmental Assessment (EA) and draft backcountry management plan (BMP) for Canyonlands National Park that included a preferred alternative establishing a permit system for ORV use and limiting the number allowed at any time. Prior to this, the Salt Creek Trail route in the park was growing in popularity with ORV drivers because it provided easy access to a popular destination in the park, Angel Arch. Because this trail through a narrow canyon runs in and out of the stream bed, and since this is the only year-round creek in the park, the NPS had determined that resource damage was occurring. However, after extensive public comment, and even though the preferred alternative in the EA called for closing a ten-mile segment of the Salt Creek Road, when the final BMP was issued on January 6, 1995, the NPS did not prohibit vehicular driving in the creek on that segment of trail. Instead it instituted a locked gate and permit system, acknowledging that impacts would continue to occur but that, by limiting the number of users, those impacts would not be sufficient to cause impairment.

The NPS was then sued by the Southern Utah Wilderness Alliance (SUWA) for violating the Organic Act prohibition on impairment of park resources and values, among other issues in the complaint. The Utah Trail Machine Association intervened in support of the NPS position, argued by the Justice Department, that the NPS Organic Act intended a "balancing" interpretation between preservation and use. On September 23, 1998, the district court entered a final order granting judgment to SUWA with respect to the ten-mile segment of the Salt Creek Trail. The judgment remanded the case to the NPS for appropriate action in accordance with the judgment and enjoined the NPS from allowing motorized vehicle travel in Salt Creek Canyon above Peekaboo Spring.

In 1999, subsequent to this district court decision, the NPS developed its newly expanded interpretation of the Organic Act's requirement to prohibit impairment of park resources. NPS professional managers, and its lawyers in the DOI solicitor's office, determined to publish a new edition of *Management Policies*, which became the 2001 edition. Section 1.4.4 of the 2001 edition states, "While Congress has given the Service the management discretion to allow certain impacts within parks, that discretion is limited by the statutory requirement (enforceable in federal courts) that the Park Service must leave park resources and values unimpaired."

In August 2000, the 10th Circuit US Court of Appeals reversed the District Court opinion, rejecting the earlier NPS interpretation of its mandate taken in the 1995 BMP, but essentially siding with the new interpretation of the Organic Act that NPS had just approved in preparation for the new edition of *Management Policies*: it distinguished between *allowable impacts* to park resources and values, and *prohibited impairment* of park resources and values.

Ironically, perhaps, NPS had closed Salt Creek to vehicles on the controversial ten-mile stretch of the stream while the planning and litigation were going on, and simultaneously began, with university partners, a series of scientific studies of the stream, including monitoring changes in water quality and stream biota. Since this ORV trail had been open to vehicles from even before the park was established in 1964, but now had been closed to such use for nearly a decade, the NPS took this opportunity to study any detectable changes. Largely as a result of these impacts revealed by the science, in June 2002 the NPS published for comment a second Environmental Assessment, including a preferred alternative that would permanently close the ten-mile section of Salt Creek to vehicles. Based on the second EA, in August 2003 the NPS published in the *Federal Register* a proposed rule to prohibit vehicle use on Salt Creek, again for public comment. A final rule was announced in the *Federal Register* in June 2004, closing the ten-mile segment of Salt Creek to vehicles. Subsequently, in August 2004 a coalition of off-road vehicle organizations sued the NPS, opposing the final rule. The coalition argued that the NPS's decision was arbitrary and capricious and violated their interpretation of the NPS Organic Act's mandate of a dual mission for NPS, to balance preservation and use.

On September 12, 2005, the District Court issued its new ruling, concluding that "the 2001 Management Policies are the type of agency decision intended to carry force of law."[32] This was not only a big win for the NPS position on Salt Creek but, for the first time in federal court rulings, NPS *Management Policies*, and the public process by which NPS developed the 2001 edition, were accepted as warranting court deference, even though the manual was not itself promulgated as a rule.

The court found that the evidence in the record amply supported the NPS's determination that motor vehicle use on Salt Creek Road would cause "impairment." The court's opinion stated, in part:

> In addition, the majority of courts that have interpreted the "no-impairment" mandate have interpreted it as placing an "overarching concern on preservation of resources." . . . The Management Policies' interpretation therefore is consistent with over twenty years of federal court decisions confirming that conservation is the predominant facet of the Organic Act. In sum, upon review of the express language of the 1916 Organic Act and its legislative history, the 1978 Amendment to the Organic Act in the Redwood Act, and its legislative history, and the interpretation of the no-impairment mandate taken by the majority of the courts, the Court was satisfied that the interpretation of the "no-impairment" mandate of the Organic Act set forth in section 1.4 of the 2001 Management Policies is permissible.[33]

JONATHAN

As a superintendent in three different national parks, I turned to *Management Policies* for an answer, a guide, and a shield, especially when I said no to the latest bad idea that came across my desk. Here are some real-world examples from my experience:

Delicate Arch with visitors, Arches National Park. Credit: Destry Jarvis.

- Would I permit scientists to use explosives on top of a park glacier to determine the thickness of the ice? *Management Policies* answer: No, the area is designated wilderness and the explosions would be heard by visitors who have an expectation of solitude and quiet.
- Would I let the state Fish and Game come into the park and develop artificial water sources for specific wildlife that were of interest to the state? *Management Policies* answer: No, without specific legislation, this would be a derogation of park values and purposes, which subverts natural processes.
- Should I allow a small lightning fire to continue to burn? *Management Policies* answer: Yes, with caveats, as fire is a natural part of the system, but it could become larger and leave the park, so it must be monitored and managed per a fire management plan.
- Would I extend the permit for an oyster farm to continue to operate in designated wilderness? *Management Policies* answer: No, commercial operations are not allowed in wilderness, and the oyster farm was causing impairment to park resources.

Across the system, park superintendents must respond frequently to complex issues and proposals, and *Management Policies* gives their decisions protection from local political pressures. While park superintendents have no authority to make a decision that is directly contradicted by policy, they can request a policy waiver that can only be issued by the NPS director. However, if the entities proposing the activity have political connections, they often pursue an overturning of the local park superintendent's policy decision by putting pressure on the director or the Secretary of the Interior.

In January 2001, George W. Bush became president of the United States and filled the position of Secretary of the Interior with Gale Norton, an attorney from the conservative Mountain States Legal Foundation. Norton was a staunch antigovernment right-winger, mentored under the tutelage of the infamous former Secretary of the Interior James Watt. Norton brought in a team of political appointees who were determined to hand the parks and public lands over to the modern-day robber barons: oil executives, big game hunters, the off-road vehicle industry, coal miners, and timber beasts. During these hostile administrations the NPS tries to lie low and stay off the radar. I once had to travel with Secretary Norton around the Northwest as she promoted the public land policies of the Bush administration. Wherever we went, we heard strong support for the NPS—even on the radio show of conservative Kirby Wilbur, who turned out to be a Civil War buff and had high praise for the park rangers at the battlefields. Sitting with Secretary Norton in the back seat of her car, driven by her Park Police protection detail, I suggested that I put together a platform of positive support for the National Parks for President Bush. Her response is burned into my memory. She turned her icy gaze on me, pointed a finger that reminded me of Cruella de Vil, and said that the NPS costs seventeen dollars an acre while the BLM costs only three dollars an acre. I said, "In all due respect, the Statue of Liberty is not a three dollar an acre property." She never spoke to me again.

I was the superintendent of Mount Rainier National Park, a much beloved park in the Northwest and a good place to hide. Norton brought on Fran Mainella, formerly director of Florida State Parks, as the NPS director. Fran was a "park person" but struggled with the complexity and breadth of the NPS. One of her assigned tasks was to cut the budget of the National Park Service, a common approach during Republican administrations. As the regional directors pointed out to her, reducing the budget meant reducing the services to the public. Her team came up with the euphemism of "service-level adjustments" to poorly mask the changes the public would see in the parks when their budgets were severely cut. In addition, her lack of understanding of the importance of the NPS *Management Policies* led to near disaster.

One of the political minions of the Department of the Interior was Paul Hoffman, a deputy assistant secretary for Fish, Wildlife and Parks. Hoffman was technically the first-line political appointee overseeing the National Park Service. Hoffman fit the mold: big moustache, big cowboy hat and boots, trophy animals on the walls of his office, and a résumé that included working for former Congressman and now Vice President Dick Cheney in his Cody, Wyoming, office. Hoffman had a particular hatred for the NPS at Yellowstone National Park. He had previously been the executive director of the Cody, Wyoming, Chamber of Commerce, and engaged on every Yellowstone issue that had an effect on the economics of the town.

Many decisions by NPS superintendents to deny an activity that was proposed by a supporter of President Bush or Secretary Norton rolled up to Hoffman for review. Consistently, the park superintendent would cite the chapter and verse of *Management Policies* for the decision. As more and more of these came to his desk, Hoffman began to edit the *Management Policies* manual in Microsoft Word, altering the intent and language of each issue that crossed his desk.

As things stood, if someone proposed installing zip lines in Yosemite National Park to the superintendent, *Management Policies* would give the superintendent the guidance to deny the activity, in that it is not dependent on the park to execute and could easily be done outside park boundaries. In addition, even if the activity was a good idea, the onus would be on the proposers to do all the analysis of potential impacts to the park at their own expense. This is a pretty good deterrent to a lot of bad ideas.

Hoffman's rewrite would have shifted that onus onto the NPS, where any new activity would be permitted until the NPS could prove, at its own expense, that there was harm. This would open up the parks to all kinds of activity that would be inappropriate and harmful to the resources. Hoffman even bought the strange interpretation of the 1916 NPS Organic Act, concocted by President Ronald Reagan's DOI Assistant Secretary Bill Horn as we have noted, that the impairment standard applied not to the resource but to visitor experience. In other words, the National Park Service could not "impair" the visitor experience by limiting what people could do in the parks! The concept that there were "inappropriate activities" in a park was foreign to these far-right political appointees and some members of Congress. When former Director Roger Kennedy was asked what would be inappropriate visitor activity, in a phrase that only the silver-tongued Kennedy could concoct, he responded "feckless merriment."

When a Republican-led Congress threatened to begin hearings on revising the NPS Organic Act, Director Mainella offered up revisions to *Management Policies* instead. She said that Deputy Assistant Secretary Paul Hoffman was already working on it. By this time, I had moved from Mount Rainier to become the new regional director of the Pacific West Region, supervising fifty-eight national parks in California, Washington, Idaho, Nevada, Oregon, Hawaii, and the Pacific Islands. Director Mainella allowed only her trusted team in Washington and the seven regional directors to see Hoffman's revisions.

When I read his draft, I was livid, as he was gutting the very framework upon which park stewardship was built. In a classic mistake, the "track-changes" version of Hoffman's rewrite was sent to us so that we could actually see each of his edits, with Microsoft Word identifying Hoffman's computer by name. On August 5, 2005, I sent an official memorandum to the deputy director with my assessment:

> This draft document is the largest departure from the core values of the National Park System in its history, posing a threat to the integrity of the entire system. These policies, if implemented as proposed, will weaken the management of parks, threaten critical resources, reduce the visitor's ability to experience nature in a form as close as possible to natural, and move the National Park System from its position as the conservation leader in the world to just another average public land manager. If our goal is mediocrity, then this document is our road guide.[34]

My memo went on to detail the specific transgressions: making all park decisions with visitor use as the priority, creating a hierarchy of greater and lesser parks, eliminating science and scholarly research, and opening up the parks to just about anything anyone wanted to do unless the NPS could prove the activity would be detrimental to the resource.

I was the only one of the internal reviewers to put my objections in an official memo to Washington. Director Mainella was furious, and I was summoned to Washington and screamed at and berated by her and threatened with discipline or removal by her loyal deputy Steve Martin.

When I signed up to work for the NPS, I joined to support the national parks and the legislative mission that these special places were to be conserved unimpaired for the people and future generations. I was not going to sit back and watch them dismantled by this or any administration. Working with the conservation community, including my brother Destry, we decided that while management policies were a pretty dry story in the press, Paul Hoffman, the former staffer to Vice President Dick Cheney and "all hat and no cows," was a pretty good target. Together on weekly calls, we launched a campaign to target Hoffman and his political connections. As the media heat increased, Hoffman became increasingly furious, lashing out at me and other career employees. Director Mainella defended him and his "minor changes" in a document she did not understand and whose importance she did not value. At some point and by someone—not me—the Hoffman draft was leaked to the conservation community and the press, adding fuel to the media fire.

Deputy Director Martin, under pressure to resolve this issue, organized a small gathering of selected individuals to negotiate the language of the changes directly with Hoffman. I was initially banned from that meeting because of my official memo criticizing the changes, but I was offered the opportunity to send a surrogate. I chose Bill Paleck, then superintendent of North Cascades National Park. I chose Bill because he was very smart and articulate but, even more importantly, I had not only witnessed but experienced Bill's sly ability to move a group in his intended direction. Big as a house and soft spoken, he used his size and his intellect in a combination of intimidation and coherent argument to win. Bill was also a good writer and he would offer a revision to the Hoffman edits that restored the policy in subtle ways. This approach was extraordinarily effective in the negotiations with Paul Hoffman.

In the meantime, the conservation community continued its media assault on Hoffman and Director Mainella, suggesting that the rewriting of the policy by this team could not be trusted unless I was there. Finally, Martin relented and allowed me to participate. Over several days, we were able to fix many of the issues in the document, but Hoffman dug in on some of the more problematic ones, as he was still fixated on changing the fundamental stewardship of the NPS. The next version of the draft was still terrible and there were very tight controls on its release. We were all threatened by Director Mainella with severe punishment if we leaked the final draft.

Things in the Department of the Interior had become rather messy, as investigations of lobbyist Jack Abramoff and his payments for political favors were underway, implicating some of the political appointees in the department. Deputy Secretary Steve Griles was found guilty by the Inspector General, forgiven by Secretary Norton, and then he resigned but wound up in federal prison for lying to Congress. Secretary Norton followed soon with a hasty exit, to "spend more time with her family," the standard euphemism for getting out of Washington when the kitchen gets too hot.

The new Secretary was Dirk Kempthorne, a charismatic Idaho politician who had served as mayor of Boise and governor and US senator for Idaho. Kempthorne inherited this mess and began to bring in his own people to clean it up. One of the first to go was Director Mainella, who was unceremoniously tossed and replaced with a career NPS administrator, Mary Bomar. Deputy Director Steve Martin saw the handwriting on the wall, forced out Joe Alston, the highly competent and well-regarded superintendent of the Grand Canyon, and gave himself that job.

Kempthorne nominated Lynn Scarlett, then DOI assistant secretary for Management and Budget, to fill the deputy secretary vacancy left by Griles, but there was a hold on her confirmation in the Senate by Colorado Senator Ken Salazar. In a series of quick events I will never forget, I received a phone call from Craig Manson, the assistant secretary for Fish, Wildlife and Parks (Hoffman's boss). Manson said "Jon, the confirmation of Lynn Scarlett hangs in the balance of you doing what I instruct. And if ever asked, this phone call never happened. You are to go to a Kinko's or other business that has a fax machine, but not from your office, and fax your infamous Management Policy memo to the attention of Steve Black, in the office of Senator Ken Salazar.[35] Make sure you call him so that he is by the machine and can pick it up. You are then to offer to come to Washington and meet with them to discuss your memo on *Management Policies*." I said yes sir, and did exactly as he asked. Steve Black asked that I come to DC and I met him in the office of Senator Salazar and we went over the issues I had raised in my memo. Soon thereafter, Senator Salazar used his hold to extract a demand on the department to drop the *Management Policies* changes in exchange for releasing the hold on Scarlett.

Within a few days, Hoffman was removed as deputy assistant secretary over the NPS and reassigned to the DOI budget office, and *Management Policies* was restored. There were some changes in the document so that the department could save face, but all of Hoffman's revisions were removed. Not long afterward, Secretary Kempthorne came out to California and I spent the day with him in Golden Gate National Recreation Area. I liked him a lot, and his easygoing manner led me to thank him for ending the debacle over *Management Policies*. He laughed and said "yes, but it upset my base"! A few months later, I ran into Hoffman in the hall of the DOI and suggested I send him a copy of the new *Management Policies*. He laughed and said he would autograph one for me, "except they took out all my changes"!

The NPS *Management Policies* have been and will be revised and expanded as new activities emerge: the authors of the current version (2006) had never seen a Segway, a battery-powered skateboard, or a remotely operated personal drone. Revisions take time, so between editions of *Management Policies* there are Director's Orders: short and timely, these policies give guidance to the field on topics from emerging technology to climate change.

During my tenure as regional director for the Pacific West, the impacts of anthropogenic climate change loomed larger and larger. Through a series of workshops that I organized between 2002 and 2005, park superintendents, resource managers, and scientists discussed the changes we were already seeing in national parks from climate change, what we could expect in the future, and what was the role of the National Park Service. We noted that fires were burning longer and hotter and the vegetation that came back was different. Rain and snow patterns were changing in the Pacific Northwest, resulting in catastrophic floods and glacial outwash. We concluded that the

NPS had essential roles to play in monitoring the changes, adapting to a new normal, and helping the public understand the impacts of climate change. It was clear to me this needed to be addressed in policy guidance for parks, but we were still in the George W. Bush administration and even speaking about climate change was strongly discouraged.

When I became NPS director in 2009, the Obama administration wanted to become a leader in addressing climate change, and I was ready to initiate a climate response strategy. Having endured the multiyear *Management Policies* rewrite under the previous administration, I knew we would have to handle this through a Director's Order.

Many of the fundamental policies of the national parks as they relate to natural resources stem from the 1963 report prepared by A. Starker Leopold, the son of conservation writer Aldo Leopold (*A Sand County Almanac*). Secretary of the Interior Stewart Udall commissioned Leopold, then a professor at the University of California, Berkeley, to investigate and recommend a new management paradigm for wildlife in the national parks, fueled by a dramatic die-off of elk in Grand Teton National Park. The subsequent Leopold Report, as it became known, recommended that the National Park Service embrace a policy of allowing nature to thrive, to study both predators and prey, and even to restore ecosystem processes like wildfire.

Leopold envisioned that parks would represent "vignettes of primitive America," re-creating and maintaining natural settings as they were before colonial expansions. At first, the NPS was reluctant to embrace this radical approach, since it had been practicing wildlife and forestry management as taught in the colleges and universities at the time: put out fires, kill predators like wolves and bears, feed popular and charismatic wildlife like elk, stock lakes and streams with fish, spray for insects, and cut down trees that might have disease. But over time, the NPS began to embrace and to actively practice this new approach. Naturally ignited fires were allowed to burn in most circumstances, predators like wolves were reintroduced, native forest diseases and insects were allowed to continue, and fish stocking stopped.

This new framework made its way into *Management Policies* and was now not only embraced by the professionals in the NPS but also by the public and the courts. It had worked pretty well for most of fifty years, but there were some gaps in the Leopold policy. For one, he had ignored that Native Americans had been manipulating the American landscape with fire and other activities for thousands of years. And, of course, he did not anticipate that a changing climate due to a warming planet would begin to impact park resources. From our scientists, we were learning that the snowpack in the Rockies, Cascades, and Sierra mountains was dwindling and holding less water. Hurricanes were becoming more intense and frequent, impacting our coastal parks. Wildfire season was now twelve months long and some fires were almost uncontrollable. Warmer winters allowed beetles to survive and decimate forests. And wildlife was on the move, forced to seek refuge in new habitats as local conditions altered. These were not just anecdotal observations. Recent research, such as the University of California, Berkeley, Grinnell Resurvey,[36] was scientifically documenting the changes. Since these effects were human-caused, technically the NPS should have begun taking action to mitigate their consequences. But policy did not guide managers as to what actions were appropriate.

I created a new Climate Response program within the NPS, hiring top scientists and assigning the task of developing a national response strategy and a framework for individual national parks to develop their specific responses. But I knew we needed new policy and, to have maximum credibility, it must stem from an outside, independent set of recommendations. I turned again to the National Park System Advisory Board to lead the effort to "revisit Leopold." I assigned the science advisor to the director, Dr. Gary Machlis, to coordinate, and I asked the former director of the National Science Foundation, Rita Colwell, to lead a team. Every person we asked to participate, without compensation, agreed to join the team of distinguished scientists and scholars representing a broad range of disciplines.

For the next twelve months, under the umbrella of the National Park System Advisory Board, the team traveled to parks, met with staff and scientists, and deliberated the goals of natural and cultural resource management in the national park system in light of these new challenges. The team was tasked to answer three questions:

1. What should be the *goals* of natural resource management in the national park system?
2. What general *policies* for resource management are necessary to achieve these goals?
3. What actions are required to implement these policies?

On August 25, 2012, the team submitted its report, concisely constrained to the same twenty-three pages as the original Leopold Report. Their conclusion:

The overarching goal of NPS resource management should be to steward NPS resources for continuous change that is not yet fully understood, in order to preserve ecological integrity and cultural and historical authenticity, provide visitors with transformative experiences, and form the core of a national conservation land- and seascape.[37]

The report also introduced a new concept, the "precautionary principle":

> Because ecological and cultural systems are complex, continuously chang-
> ing and not fully understood, NPS managers and decision makers will need
> to embrace more fully the precautionary principle as an operating guide. Its
> standard is conservative in allowing actions and activities that may heighten
> impairment of park resources and consistent in avoiding actions and activ-
> ities that may irreversibly impact park resources and systems. The precau-
> tionary principle requires that stewardship decisions reflect science-informed
> prudence and restraint.[38]

An independent report does not constitute policy for the NPS, so now the
task was to convert these recommendations into something usable in the
field. I also knew that the existing approach to policy—the original Leopold
paradigm—was so ingrained in the field that any change would have to be
informally and formally reviewed by the rank and file as well as our outside
observers, critics, and advocates. In 2015 I initially sent out a policy memorandum to provide interim guidance. Then we began the arduous task of
writing official policy.

Also in 2015 a new team was assembled and given one year to develop the
Director's Order. This team came from the NPS career ranks and was led by
two of the most respected senior managers: Midwest Regional Director Mike
Reynolds and Golden Gate NRA Superintendent Chris Lehnertz. The team
was to draft Director's Order 100 and finalize it in 2016, coinciding with the
100th birthday of the NPS. Dr. Machlis, serving as coordinator, provided the
continuity between the "Revisit Leopold" team and the new "DO 100" team.
They worked diligently for a year and subjected the draft to rigorous review
both internally and externally. It was an ambitious process, requiring nearly
simultaneous review and rewrite.

On December 20, 2016, I signed the new Director's Order 100, giving the
field clear guidance on how to manage national parks under the threat of a
warming planet, rising seas, and a turbulent climate. For decision making,
the DO established the following framework:

- *Best available science and scholarship* is defined as up-to-date and rigorous in
  method, mindful of limitations, peer-reviewed, and delivered at the appropriate time in the decision-making process in ways that allow managers to
  apply its findings.
- *Accurate fidelity to the law* means that the NPS decision-making process
  must adhere with precision to the law, be mindful of legislative intent, and
  consistently and transparently follow public policy and regulations.

- *Long-term public interest* emerges from the NPS mission, the expert judgment of NPS professionals, and an evolving understanding of public wants and needs across multiple and inclusive perspectives.

Director's Order 100 required that all park superintendents possess scientific literacy appropriate to their positions and resource management decision-making responsibilities. The DO integrated natural and cultural resource programs and functions wherever beneficial to resource stewardship. And it expanded internal and external partnerships to achieve the policy objectives of serving as core to a national and international network of protected lands and waters.

But, of course, Director's Order 100 was signed in the window between election and inauguration, with Donald J. Trump to become president in January of 2017. He soon populated the department with developers, government haters, public land privatizers, lobbyists for the oil and gas industry, and appointees returning from the Bush years who were wise to the NPS and its frequent resistance to political pressures.

With a stroke of the pen, ordered by new Secretary of the Interior Ryan Zinke (who arrived on an NPS horse for his first day of work), Director's Order 100 was rescinded. Mike Reynolds, now acting director, who had led the team that wrote DO 100, was forced to inform the field that the DO was rescinded and no longer in effect. While the "official" version of the new policy has been rescinded, I know that the dedicated employees of the NPS have copies in their desk drawers and are taking actions to steward their parks during these times of uncertainty.

Over the course of the Obama administration, the citizens I had asked to serve on the National Park System Advisory Board produced a significant legacy. They had led the effort behind Revisit Leopold, supported new National Historic Landmarks to better represent the diversity of the nation, fostered a new approach to the role of the NPS in public education, and advanced the NPS role in philanthropy. These distinguished individuals had volunteered hundreds of hours, traveled to parks and to Washington on multiple occasions, and were deeply knowledgeable about the mission and goals of the NPS. They requested to meet with Secretary Zinke and were repeatedly rebuffed throughout 2017. Frustrated, in January of 2018, ten of the twelve resigned in protest. In a letter to Secretary Zinke, Advisory Board Chair and former Governor of Alaska Tony Knowles stated:

For the last year we have stood by waiting for the chance to meet and continue the partnership between the NPSAB and the DOI as prescribed by law. We understand the complexity of transition but our requests to engage have been ignored and the matters on which we wanted to brief the new Department team are clearly not part of its agenda. I wish the National Park System and Service well and will always be dedicated to their success. However, from all of the events of this past year I have a profound concern that the mission of stewardship, protection, and advancement of our National Parks has been set aside. I hope that future actions of the Department of Interior demonstrate that this is not the case.[39]

Park Policy Wonk and Professional Park Manager—We Conclude That Policy Provides Essential Guidance When It Isn't Politically Manipulated

One of our old friends in the BLM once said that the NPS will work really, really hard to say no to just about anything while the BLM will work really, really hard to say yes to just about anything. That is why Burning Man takes place on BLM and not NPS lands. To us, that is what makes it a great system of parks and public lands. The efforts of hostile administrations to force the NPS to accommodate all kinds of crazy things that are more appropriate elsewhere continues to be thwarted by *Management Policies*.

The infuriating saga of repeated attempts to revise the policy framework of the National Park Service and the defensive battles to protect it by ourselves and many others demonstrate how dangerous it is to leave the stewardship of our national parks subject to the ideology of hostile political appointees. Although most of these battles have been won for preservation, the energy expended could have been used to do something positive for the stewardship of our national parks and for the American people. During the Trump administration, Secretary David Bernhardt continued the attacks on the NPS begun during his eight years during the George W. Bush administration as solicitor for the DOI. Now wiser to the ability of the agency to resist, and well aware of both numerous court rulings and broad public opinion that favor the NPS professional interpretation of the Organic Act, he ignored new changes to *Management Policies* and focused instead on the career employees, forcing many into retirement or silence through threats and intimidation. He intentionally kept the NPS director's position an unstable assignment, with four individuals filling in as acting directors over Trump's four-year term, two of them purely political appointees.

As only the latest indicator of the distain with which the Trump and other recent Republican administrations have held career professional NPS and other federal employees, in October 2020, President Trump issued an executive order instructing the most significant restructuring of the career civil service in a century. Specifically, his order would move all career employees with a role in policy making, such as park superintendents and headquarters staff, into a category of employees called "excepted service" that allows their swift removal from public service by political appointees when they disagree with the administration's policy direction. With the election of Joe Biden as president, this drastic policy was rescinded, but it sent shivers through the career ranks. With the confirmation of Deb Haaland as DOI Secretary, the first Native American to hold that position, the political pendulum has swung back to someone who will support and sustain the professional staff of NPS and not repudiate, intimidate, and reprimand them for adhering to the NPS mission.

For over a hundred years, Congress and the courts have largely acted to protect the natural and cultural heritage of our nation, held in stewardship by the National Park Service. The NPS, in its responsibility, has crafted policies that guide the decision makers in their solemn duties. But with every change in administration, from Democratic to Republican, comes a new assault on these policies and the resources they protect. The only real solution is to give the NPS the independence it must have to achieve its mission.

FOUR

USING THE BEST
AVAILABLE SCIENCE

> The very foundation upon which the National Park Service is built [is]
> the preservation of the native values of wilderness life. For it is this
> ideal above all else which differentiates this service from its sister ser-
> vices in government.[1]

Through much of its first hundred years, NPS field managers as well as directors were focused on building up a constituency for the parks through
visitor experiences and associated educational writings to reach ever more
of the American people. These early leaders, like Directors Mather and Albright, began by actively recruiting the railroad and stagecoach companies to
promote park visitation. With the advent of the private automobile, and especially after World War II, visitation soared, and expansive service facilities
and roads were added to the parks. Most early parks, in remote corners of the
Western public lands, were not being encroached upon by mining and timber cutting at their boundaries, or by growth in nearby gateway communities,
as they are today. Air and water pollution in these remote locations was not
yet a serious problem, and native wildlife, especially the "charismatic mega-
fauna" like elk and deer, birds and beaver, were more abundant and viewable
by visitors than elsewhere.

Essentially the parks' ecological integrity was not a dominant management
issue. By the 1960s and 1970s, this situation was rapidly changing—air and
water pollution were serious, wildlife numbers were in decline, toxic wastes
were washing in, acid rain was falling, and soaring visitation was in serious

need of active management. The service needed a change in its management priorities as well as in its internal culture, which tended to put "use" ahead of "preservation."

From the inception of the national parks, park superintendents often have had neither sufficient scientific data to inform decisions nor the scientifically trained staff to interpret what data they had. Politics did not enter into the policy decision-making process in any meaningful way when it came to application of sciences in the parks, but there was a basic lack of research and professional resources management to inform management decisions. However, beginning in the 1970s, and accelerating in the 1980s, abrupt natural resources policy changes resulted from changes in administration and especially their Secretaries of the Interior.

Several Secretaries—James Watt, Gale Norton, and David Bernhardt in particular—have actively sought to suppress, ignore, or rewrite science to support their politically driven resource development or ideological philosophies. Secretary Watt was known for stating, during his morning prayer meetings, that there was no need for conservation, since Jesus was returning soon and the world of humans on Earth would end. In 2019, Secretary Bernhardt authorized the practice that allowed hunters in Alaskan national preserves to kill female bears and their cubs in their winter dens, based on the unscientific theory that such killing would improve caribou shooting for sport hunters. While some would argue this is a traditional practice of the Native Alaskans, it was done only during times of starvation when a native hunter would enter a bear den with torch and spear. That is a far cry from a modern hunter outfitted with a flashlight and a gun.

As the new NPS began its mission after 1916, its earliest scientific studies, though few and far between, were basic field science. For example, the 1918 *Wild Animals of Glacier National Park*, by Vernon and Florence Merriam Bailey, contains no reference to policies or how the park should be managed. In fact, Vernon Bailey was the chief field naturalist of the Bureau of Biological Survey at the US Department of Agriculture—there were *no* scientists at the US Department of the Interior involved in national park management.[2]

The NPS hired its own first scientist, George Melendez Wright, in 1927. His main job was to prepare a series of scientific studies of park wildlife, known as the *Fauna Series*.[3] Soon after joining the NPS, Wright used his personal wealth to fund a small group of scientists at the University of California, Berkeley, to prepare this set of park biological studies intended to give the NPS a sense of policy direction for managing native plants and animals. Wright began to advocate a new policy approach with his studies— that parks should manage and protect the full variety of native wildlife, not just the charismatic species popular with visitors.

When the first volumes of the *Fauna Series* were published in 1932 and 1933, NPS leadership began to recognize that science-based conservation decision-making was important to do, but it did not adopt strict science-based, decision-making policies as Wright had recommended until some forty-five years later.

Fauna No. 1, *Fauna of the National Parks of the United States*, included a set of proposed management policies that were never officially adopted by NPS. Nevertheless it foretold the need for management decision-making based on the application of whatever biological knowledge was available. Fauna No. 1 stated in part,

> Relative to areas and boundaries—That each park shall contain within itself the year round habitats of all species belonging to the native resident fauna. . . . Relative to management—That no management measure or other interference with biotic relationships shall be undertaken prior to a properly conducted investigation. . . . That the rare predators shall be considered special charges of the national parks in proportion that they are persecuted everywhere else. Relative to relations between animals and visitors—That presentation of the animal life of the parks to the public shall be a wholly natural one. . . . That no animal shall be encouraged to become dependent upon man for its support. . . . Relative to faunal investigations—That a complete faunal investigation . . . shall be made in each park at the earliest possible date. . . . That each park shall develop . . . a personnel . . . trained in the handling of wild-life problems, and who will be assisted by the field staff appointed to carry out the faunal program of the Service.

NPS policy begins with the "conserve . . . unimpaired" language of the 1916 Organic Act and was amplified by the 1918 Lane Letter's three principles. These foundational principles charge NPS to:

- conserve the parks unimpaired for future generations,
- manage parks for public use and interest, and
- ensure that the national interest dictates all decisions.

This fairly simple set of policies was used by the NPS for decades, and the far more detailed policy recommendations proposed by George Wright were largely ignored. But Wright's suggested policies were predictive of the many serious management problems that would beset the NPS for much of the remainder of the twentieth century.

During the Great Depression years of the 1930s, Wright, again using his own private fortune but now supplemented by funds allocated to Civilian Conservation Corps (CCC) programs in the parks, recruited and trained as many as twenty-three wildlife biologists. These men (and they were all men at that time) began to have a positive impact on knowledge of park natural resources.

The service even (briefly) adopted a policy requiring that the biologists review all park management and development projects that could impact park wildlife. Tragically, Wright was killed in a car wreck in 1936 and, by 1939, as a result of the Roosevelt government reorganization, the entire NPS wildlife division, including its field personnel, was transferred to the US Fish and Wildlife Service (USFWS) as the Park Wildlife Division. This action had the unintended consequence of setting back NPS research and resource management by almost two decades. The NPS needs research to inform park managers on actions to take or not take—actions directly related to managing the park. Scientists embedded in the FWS could not be expected to do that sort of management research; their mission is entirely different. While there may be some overlap regarding endangered species or migratory birds, the FWS is much more focused on species, while the NPS manages for ecosystems.

But the idea that the NPS could revert to managing the national parks with its earlier, simple policies began to change rapidly in the 1960s. An initial push toward more scientific work came as a result of the First World Conference on National Parks, held in Seattle in 1962. A fifteen-member, eight-nation committee at that conference produced a report, *Management of National Parks and Equivalent Areas*, that concluded "few of the world's parks are large enough to be in fact self-regulatory ecological units; rather most are ecological islands subject to direct or indirect modification by activities and conditions in the surrounding areas." It further determined that "management based on scientific research is, therefore, not only desirable but often essential to maintain some biotic communities in accordance with the conservation of a national park."

It was in large part as a result of this international conference that DOI Secretary Stewart Udall convened the first NPS Advisory Board on Wildlife Management, chaired by Dr. A. Starker Leopold. This advisory board presented the aforementioned Leopold Report on March 4, 1963, which noted that "on the whole, there was little major change in the Park Service practice of wildlife management during the first 40 years of its existence." And at that point, the NPS had been deeply engaged for nearly a decade in its most extensive and rapid facility construction period, Mission 66, building miles of road and hundreds of buildings, including visitor centers and staff housing. Until the Leopold Report, the focus of NPS leadership and policy was on visitors, tourism, and recreation, largely on the (unsubstantiated) theory that nature would take care of itself but visitors needed facilities and services.

By the early 1960s, wolves and mountain lions were gone due to an active shooting campaign, and Yellowstone was overrun with and overgrazed by elk. The NPS responded by shooting more than 4,000 elk in one winter alone. What began to change all that was the public outcry over the ongoing NPS policy that favored the "good" species (e.g., elk) over the "bad" species (e.g., wolves). In Yellowstone, for example, the NPS (and all other federal land management agencies) had pursued for decades a policy of shooting and trapping predators, especially wolves and mountain lions. The goal was to increase the population of elk, which visitors loved to view and others loved to hunt as they moved outside park boundaries.

Secretary Udall appointed Leopold as NPS chief scientist, the first such title, even though he was actually only a consultant. Deputy Chief Scientist Dr. Robert Linn, a career NPS employee, reported to Leopold. At least one key senator, Lee Metcalf (D-MT), objected to this odd and inadequate arrangement and made his concerns known in a series of letters in 1967 that were included in the annual Appropriation Committee report. Metcalf said:

> Research is basic to proper administration of these [park] lands, and the Park Service's research program is inadequate. It is inadequate in terms of exploring plant, animal, geological and other natural aspects of the parks and intensively studying the public-use situations that have arisen and will be compounded in the future. There must be better understanding of visitation, transportation, facilities and services, and other park-related matters if units of the park system are to remain natural preserves within the meaning of the Act which set up the Park Service in 1916. Of the total budget request . . . for the Park Service, only 1.8% is allocated to the research on which sound administration is based.[4]

The central thesis of the Leopold Report was:

> As a primary goal, we would recommend that the biotic associations within each park be maintained, or where necessary recreated, as nearly as possible in the condition that prevailed when the area was first visited by the white man. A national park should represent a vignette of primitive America . . .[5]

As time passed, this goal came under fire. Sport hunting organizations claimed it told NPS to take a "hands-off" approach to resources management, that NPS should let nature take its course, rather than engage hunters to control animal populations, as they wanted done. Other critics insisted that the Leopold Report ignored the fact that all of the lands and waters now in national parks had experienced centuries of occupancy and use by Native Americans and thus could not be "vignettes of primitive America." But, far from recommending a hands-off approach to park management, the Leopold Report intended that the NPS take a more proactive role in understanding its natural resources, through science. It was not until later that the NPS recognized that indigenous peoples were an integral part of that primitive America. The goal now was to manage the full diversity of native biota, or restore it where it had been extirpated.

The Leopold Report of 1963 was followed closely by the Robbins Report of 1964. At the same time that Secretary Udall appointed the Leopold Committee and tasked it to apply science principles to park management, he also asked the National Academy of Sciences to appoint a committee to advise the NPS on basic research needs. It was chaired by Dr. William J. Robbins, a botanist, fellow of the National Academy, and head of the New York Botanical Garden. The observations of the Robbins Report were scathing:

> An examination of natural history research in the National Park Service shows that it has been only incipient, consisting of many reports, numerous recommendations, vacillations in policy and little action. Research by the National Park Service has lacked continuity, coordination and depth. It has been marked by expediency rather than by long-term considerations. It has in general lacked direction, has been fragmented between divisions and branches, has been applied piecemeal, has suffered because of a failure to recognize the distinctions between research and administrative decision-making, and has failed to ensure the implementation of the results of research in operational management.[6]

Over the course of the next ten years, the NPS began to get a better model for science in the parks underway but did not always match it by applying this scientific knowledge to park management in an effective manner. A chief scientist position in the office of the director was established, and cooperative relationships were developed with various research universities that could direct faculty and graduate students to undertake needed park research. And in the late 1960s and early 1970s, the NPS officially dropped its policy of removing predators from parks, stopped artificial feeding of elk or deer, and closed garbage dumps in Glacier and Yellowstone as bear-feeding visitor attractions. The NPS also during this period began to restore natural processes such as fire into parks such as Yosemite.

Changes occurred in the broader society as well as in specific park policies. Congress enacted a series of powerful new environmental laws that added to the complexity of park management but intended better protection of park resources. Included in the plethora of new laws were the Wilderness Act of 1964, the National Historic Preservation Act of 1966, the Wild and

Scenic Rivers Act of 1968, the National Environmental Policy Act of 1970, the Clean Air Act of 1970, the Water Pollution Control Act of 1972, and the Archeological Resources Protection Act of 1979.

DESTRY

Successful enactment of the 1977 amendments to the Clean Air Act was one of my first big lobbying efforts at NPCA. This major set of changes included a new legal standard, "prevention of significant deterioration" (PSD) of air quality in already clean air areas, specifically including national parks and monuments. The act defined standards for three classifications of PSD air quality: Class I allows the least additional air pollutants, Class II allows a modest increase, and Class III allows more, though still not allowing "significant" deterioration of air quality. All existing national parks were automatically in Class I, which tightly restrained any increase in air pollution that would affect them. Class II status, which allows well-controlled new facilities that emit pollutants, was applied to existing national monuments and any future national parks designated by Congress. The law also allows the NPS to propose

Yellowstone National Park, atop Mount Washburn, one of three 360-degree integral vistas. Credit: Destry Jarvis.

redesignating any national monument as Class I administratively, but only with state approval. The law encouraged states to endorse reclassifying new national parks as Class I, but no state has done so since 1977. Importantly, mandatory Class I and Class II areas cannot be downgraded to Class III.

My work for the NPCA especially focused on a provision of the act that required protection of *visibility*, with national parks specifically in mind, so that the air quality of scenic vistas is preserved or restored. Although regulation of air pollution is the purview of the Environmental Protection Agency, importantly, this act directed the "federal land manager"—the NPS in this case—to make the critical determination about whether new air pollution sources would harm the "air quality related values" of the national parks. Thus, for the first time in law, the NPS was given explicit authority to develop the science to determine impacts on park resources arising *outside* park boundaries. NPS professional staff proceeded with a service-wide survey of the national parks seeking to identify "integral vistas," those viewpoints most popular with park visitors, where the protection of visibility was essential to maintaining high-quality visitor experiences. For most of the identified integral vistas, NPS staff identified both the specific place and the specific compass angle where clear skies are most in need of protection. In only two cases did the NPS staff decide that visibility protection for a 360-degree vista was necessary: atop Mount Washburn in Yellowstone and atop Cadillac Mountain in Acadia.

Of course, all this data analysis and public communication took time, and by 1981, as the Reagan administration began, met stiff opposition, both from DOI and EPA political leadership. The idea of the NPS officially identifying integral vistas was killed and has not had political support since. Nevertheless, NPS technical staff have continued to perfect the science needed to monitor and analyze air quality, especially visibility, across the system. It is also true that nationally air quality has improved in the more than forty years since enactment of the Clean Air Act Amendments of 1977. A new administration, or an independent NPS, could readily revive and fully justify official designation of integral vistas. Additionally, no national monument has been redesignated as Class I, and even newly designated national parks, including all of the ANILCA parks in Alaska, remain less protected, as Class II. In one especially egregious case, in 1980 Chaco Canyon National Monument was redesignated as Chaco Culture National Historic Park with the specific intent that it then could not be administratively redesignated as a Class I park under the 1977 Clean Air Act Amendments. More recently, this change has enabled the Trump administration to propose major expansion of oil and gas leasing of the BLM lands that surround the park. These BLM lands include highly significant archeological sites that are also examples of the Chacoan culture, and both the park and BLM sites comprise an inscribed World Heritage Site—but cannot now be reclassified as Class I without an act of Congress.

Given that there was general consensus that the major threats to the ecological integrity of the parks were from activities outside of park boundaries and thus outside NPS control, the NPS was generally pleased with these new overlay laws, not least because they fully applied to actions of the US Forest Service and the Bureau of Land Management, the agencies with conflicting multiple-use missions that manage most of the public lands around park boundaries.

By the end of the 1970s, the NPS had embraced the application of these new environmental management standards for the parks and, in so doing, realized that the agency could not continue to rely solely on its cadre of generalist park rangers[7] to manage the parks. In the early 1980s the NPS began training a new generation of resource management specialists. These men and women had degrees in the sciences, but their jobs were not basic research but rather the day-to-day application of the knowledge derived from research to the management of park resources.

Beginning in the late 1970s, on behalf of the NPCA, I began a series of campaigns, congressional testimony, magazine articles, newspaper and television interviews, and agency meetings to call public attention to a wide variety of threats to the national parks. Air quality was still deteriorating, with acid rain damage widespread and pollution-obscured vistas even more so; water pollution of park streams and rivers was a serious problem for nearly every park that was not located at the headwaters of its watershed; the US Forest Service was clear-cutting forests right up to park boundaries, so much that the park boundary of Yellowstone could be seen from space due to the contrast; the BLM was issuing mining permits for its public lands contiguous with park boundaries.

It was abundantly clear that the NPS needed to gather scientific data across the system in order to have any hope of changing other agencies' policies that were impacting its resources. If the NPS could not monitor, inventory, and assess the condition, range, and populations of its biological resources inside the parks, it would not have any hope of influencing these other agencies.

In the fall of 1976, outgoing NPS Director Gary Everhardt again called on Dr. A. Starker Leopold, along with NPCA board member Dr. Durward Allen, to look into the status of the park science program. I met with Dr. Allen numerous times during their deliberations to discuss wildlife management in the parks and why adequate funding and policy support for research were badly needed. Dr. Allen's own foundational research, on the wolves and moose of Isle Royale National Park, first published in 1979 as *The Wolves of Minong*, continues to this day and is the longest running wildlife study in the system. In July 1977, this second Leopold Commission submitted its report and recommendations to the new NPS director, Bill Whalen.

Among Whalen's first actions after becoming director was to convene what became a monthly meeting known as the Unity Group, which I regularly organized. These key park conservation leaders included Michael Frome and Bob Cahn. Frome was the long-time conservation editor of *Field & Stream* magazine. Cahn won the 1969 Pulitzer Prize for National Reporting for the *Christian Science Monitor* with his series "Will Success Spoil the National Parks?" The focus of the Unity Group became these mounting threats to the parks and how the NPS should respond.

The Leopold/Allen Report's recommendations included putting a line item in the NPS budget for park science; establishing an associate director for natural science; and requiring that the technical functions of park scientists be supervised by scientists (although their administrative functions would still be supervised by park superintendents) in order to assure quality control over the results of park research and resources management. Finally, the report recommended that, at least in the larger parks, the NPS establish positions for resource management biologists, who would work closely with research-grade scientists but be supervised by park superintendents. These recommendations essentially became my advocacy marching orders at the NPCA for the next several years.

At that time, the NPS Western Region was the most progressive in building a science and professional resource management program, led by Dr. Bruce Kilgore, associate regional director for Resources Management and Planning. As a grad student, Bruce had worked with me at the NPCA for several years as assistant editor of *National Parks Magazine*. He was another of the long line of University of California, Berkeley, graduates closely associated with the NPS, including George Wright, A. Starker Leopold, and Directors Stephen Mather, Horace Albright, and Bill Mott. In 1978, Kilgore convened a regional conference on science and resources management, for which his keynote address began:

> During our first century of managing national parks, we took it upon ourselves to "play God"—because we decided which natural processes were "good" and which were "bad." In 1963, we were reminded by the Leopold Report that "playing God" was not what our mission is all about. It pointed out the folly of tinkering with natural processes, without understanding these processes. It said the NPS must recognize the enormous complexity of ecological communities and the diversity of management procedures required

to perpetuate them. It said that management without knowledge would be a dangerous policy.[8]

Over a number of months in 1977 and 1978, I and my NPCA staff conducted a survey of 203 NPS parks to assess the range and degree of threats facing them. We published the results in spring 1978 in the NPCA magazine as the "NPCA Adjacent Lands Survey: No Park Is an Island." As a direct result, House Parks Subcommittee Chair Phillip Burton and ranking member Keith Sebelius wrote to Director Whalen requesting that NPS prepare a "state of the parks" report to respond to the threats outlined by the NPCA.

Simultaneously, I had been working for some time with Roland (Ro) Wauer, NPS chief of the Division of Natural Resources in Washington, DC, to more directly engage the NPS in addressing the full range of threats to the parks. Over the course of the next two years, I worked closely with Ro as he developed the first ever "State of the Parks Report to Congress," which the NPS submitted in May 1980 to fulfill the congressional request. The report was organized to identify specific threats, the source of the threats (internal or external to the parks), and which park resources were endangered by the threats. Every park surveyed identified threats, 50 percent of them arising outside of NPS boundaries and control.

The report had a huge effect, both publicly and in Congress. Burton and Sebelius requested a follow-up report to Congress on what actions the NPS proposed to take on the compilation of threats to the parks. This second report, entitled "State of the Parks: A Report to the Congress on the Service-wide Strategy for Prevention and Mitigation of Natural and Cultural Resources Management Problems," was submitted to Congress in January 1981. Unfortunately, this report could only address actions that the NPS itself could take, and thus did not deal with half, and many of the direst, threats arising outside park boundaries. NPS did propose a budget initiative that would allow it to identify resource management needs and to prepare a resources management plan for each park.

Perhaps most significantly, the report proposed that the NPS would expand its internal training to develop a natural resource management (NRM) specialist training program. The first class of these thirty or so NRM trainees included brother Jonathan, who at that time had six years in the NPS as a ranger. After some intense work, Ro was able to convince the Office of Personnel Management that these future natural resource management specialists ought to be taken out of the ranger 025 job series and put into the professional 401 job series.

As the NPS itself was beginning to formulate its internal response to the

"State of the Parks" report by building the NRM trainee program and seeking to fund its research programs, I was working with members of Congress to seek a legislative solution. Representative John Seiberling (D-OH) and Doug Bereuter (R-NE) introduced bills that would, if enacted, establish a legal framework for "federal consistency" that required all other federal agencies to take the impacts of their proposed actions on national park resources into account before deciding on any projects. Hearings were held, favorable testimony taken, including from myself, but action stalled largely due to opposition from the Reagan administration and Secretary of the Interior James Watt.

Just months into the job of Secretary, Watt announced the "Park Restoration and Improvement Program" (PRIP), a proposed budget increase for NPS that focused entirely on park infrastructure construction and maintenance. He specifically forbade any new funds for science or resource management programs. The politically appointed new deputy director of the NPS, Mary Lou Grier, formerly CEO of a Texas concrete company, pointedly put resource management and science as the lowest of NPS priorities in the agency's fiscal year 1984 budget request.

As a consequence, except for the NRM trainee program, which under Ro's guidance proceeded under the political radar, any significant improvements in NPS science and resource management stalled for the next four years. However, Watt's extreme positions came to an abrupt end after less than two years when President Reagan fired him, apparently at First Lady Nancy Reagan's behest: she felt his many intemperate public statements embarrassed the administration.

At that point, President Reagan moved his friend Judge William Clark, then national security advisor, out of the White House and into the job of Secretary of the Interior. After Reagan's reelection, they chose William Penn Mott Jr. to be the next NPS director. Mott, who had worked for the NPS early in his career, was Reagan's former California State Parks director and also an NPCA board member at that time. Number one on his "Ten-Point Plan" of priorities for the NPS was "develop a long-range plan or strategy to better protect our natural, cultural, and recreational resources."

From his decades of professional park management experience, Mott was well aware of the serious external threats to the national parks from neighboring Forest Service and BLM lands. Among Mott's earliest presentations to Congress, at a July 1986 Senate hearing on the future of the national park system, he noted "We must seriously begin to think about the relationship of federal agencies with each other so that one agency does not take action that will harm the efforts and mission of another."[9]

Also in July 1986, I spoke at the NPS Fourth Annual Service-wide Science Conference, with these words:

In carrying out its responsibilities for resources management, the NPS faces two major hurdles, one political and the other institutional. Though both are formidable, the Service cannot hope to cope with the larger and more complex political problems unless and until it resolves the institutional problems. The institutional problems are both wide and deep, ranging from a failure to define clear career ladders for employees in professional series; lack of adequate in-service or academic training for generalists or specialists; a muddy organizational structure which does not optimize the abilities of either specialists or generalists at field, region, or national levels; 2-way mistrust and misunderstanding between managers and scientists; lack of a clear, accountable budget allocation process for prioritizing and adhering to priorities for centralized funds and a concomitant lack of essential funds for resource management in park base accounts; and a failure to rely on resource management plans as the first and most essential step in all park management decision-making. . . . The NPS continues to rely primarily on the "man for all seasons" generalist for most policy decision-making roles. I submit that the time has long-since come when the Service cannot maintain its prestige as a leader in world conservation, much less its own unique array of resources, without a fundamental shift away from seat-of-the-pants decision-making, and toward factually-based, scientifically derived policy setting and management decision-making.[10]

Addressing NPS employees internally in a column he wrote for the NPS *Courier* in August 1986, Mott said that "our task is to assure that the resources do not pay the price of our lack of knowledge and understanding and that if we err, we err on the side of preservation."

Unfortunately for Director Mott, and for these fine policies, for virtually all of his tenure he was forced into a defensive battle against his immediate supervisor, Assistant Secretary for Fish, Wildlife and Parks William P. Horn.[11] Horn came to that position with a strong bias against the preservation mission of the NPS. As recounted earlier, he made a strenuous effort over several years, despite Mott's objections and resistance, to reinterpret the Organic Act's NPS mission statement to mean that it was the visitor's recreation experience that should not be impaired by NPS management.

Mott, however, enjoyed the full support of President Reagan, such that Horn could not fire him, despite his efforts to do so. Instead, Horn took out his frustration and policy control efforts by seeking to destroy the careers of NPS senior managers; rejecting Mott's reorganization plans; denying Mott's promotions and bonuses to senior officials; ordering NPS regional directors and Senior Executive Service[12] superintendents to move or retire; and continuing to insist that recreation uses trump resource protection.

The assistant secretary sought to force Western Regional Director How-ard Chapman, a forty-year career NPS senior manager, to sign a statement that NPS would halt land acquisitions for Yosemite. Chapman also raised the ire of the assistant secretary over his push to halt scenic air tour flights be-low the rim of the Grand Canyon. In the end, Chapman was forced to retire as Western regional director as retaliation for testimony in Congress against Horn's policy and personnel changes. Chapman had said, "The Administra-tion's political appointees are dismembering the professional capability of the National Park Service at the national, regional and park levels."[13]

Despite these pressures, Mott continued to push for resource manage-ment reforms. His special task force appointed for this purpose rendered its report in March 1987, entitled "The Role of the National Park Service in Protecting Biological Diversity." The task force was chaired by Christine Schonewald-Cox, with the final report largely written by *Park Science* editor Jean Matthews. This concise eighty-page report concluded with ten recom-mendations. Four of the key ones were:

- One of the primary NPS missions should be to maintain and restore native biological diversity;
- Conservation of biological diversity must become the central and overriding principle for organizing management and administration of NPS natural parks and zones;
- Because parks are too small, scattered, and otherwise inadequate to achieve this goal, cooperation between NPS and federal, state, local, and nongovern-mental organizations is mandatory;
- This mission is so fundamental that science training and translation within the service and interpretation to the public are mandatory.[14]

While all these reports and recommendations were pouring into the NPS, Director Mott was still feuding with Assistant Secretary Horn over both policy and personnel. Horn ordered three NPS career associate directors and several more regional directors transferred to lesser positions. Dr. Richard Briceland was removed as NPS associate director for Natural Resources and replaced with Dr. Eugene Hester, then the head of Fisheries Research at the US Fish and Wildlife Service. The intent of these moves was to place individuals Horn could control into key NPS senior management positions—and to frustrate Mott in the hope he would resign. He did not.

In February 1988, the NPCA published volume 2 of its nine-volume *Na-tional Park System Plan*, entitled *Research in the Park: An Assessment of Needs*, largely prepared by two of my NPCA staff, Brien Culhane and Kristen Bevi-netto. We surveyed some 400 researchers in national parks and at various universities with a detailed questionnaire, and we followed that with a workshop that included both scientists and managers in each of the NPS's ten regions. The report concluded with thirty-eight recommendations, the first of which was that Congress should enact a specific legislative mandate for NPS research. The second was that the NPS budget should include a separate line-item request for science and resources management totaling 10 percent of the NPS operating budget.

The NPCA's nine-volume *National Park System Plan,* which I conceived, developed funding for, and assembled staff to prepare, was a three-year effort, funded by a grant from the A. W. Mellon Foundation. These 2,000 pages covered every aspect of the NPS and the park system, evaluating and making recommendations for everything from park boundaries to revision of the planning process. Director Mott designated Assistant Director Boyd Evison to lead an internal team to evaluate the NPCA recommendations and begin adopting or incorporating them into ongoing internal actions.

Paralleling this effort, and in close coordination with Director Mott, the NPCA appointed a Commission on Research and Resource Management Policy in the National Park System. Chaired by Dr. John Gordon, dean of the Yale School of Forestry and Environmental Studies, it included sixteen other distinguished science and policy experts, covering both natural and cultural fields. Most of the staff work and report drafting was done by Dave Simon from my staff at NPCA.[15]

The commission's final 1989 report concluded:

The public is growing more aware of a threatened world biosphere. Today the world is threatened by global climate change, acid rain, nuclear waste disposal, stratospheric ozone depletion, species extinctions, and the loss of irreplaceable buildings and artifacts that define our past. Combined, these problems present the greatest challenge, and the greatest opportunity, the National Park Service has ever faced. The challenge is that park resources are severely threatened. The opportunity lies in the possibility of finding new dimensions of value for the parks. The relatively unaltered ecosystems of the parks can provide invaluable information and service as an early warning system for many types of change. The inspiring places of our history can serve as metaphors for the diversity of America, and help sustain national unity. Only then can the National Park System continue to serve the nation as the preeminent link between our past and our future.[16]

Reports and recommendations for the NPS science and resource management program continued to pile up through the early 1990s. In 1991, the NPS convened a national conference to commemorate its seventy-fifth anniversary, in which I participated fully. This conference resulted in a lengthy set of recommendations for the agency's future, including ones for expanding science and resources management. This report, known as the Vail Agenda (after the conference location in Colorado), enjoyed enough broad support from within the NPS that its recommendations continued to resonate for several years. The vision for the future offered by the Vail Agenda included six strategic objectives, one of which, "Science and Research," asserted that "the National Park Service must engage in a sustained and integrated program of natural, cultural and social science resource management and research aimed at acquiring and using the information needed to manage and protect park resources."[17]

In 1993, the National Research Council of the National Academy of Sciences set up a Committee on Improving the Science and Technology Programs of the National Park Service, chaired by Dr. Paul Risser. Dr. Risser correctly concluded that, up to that time, there had been "an abysmal lack of response" from the NPS to the recurrent recommendations made in prior studies and reports regarding the urgent need for expanded science and resources management. Like other reports before it, the Risser Committee recommended "legislation to establish the explicit authority, mission, and objectives of a National Park Service science program."[18]

My friend and close park colleague Bob Cahn wrote a scathing piece for *Environment* magazine that year based on the Risser Committee's work, noting that many regional directors, superintendents, and protection rangers consider the time and money spent on science to be an "imposition on their responsibility to take care of the ever-increasing throng of visitors." Bob also observed that "the park service science program is unnecessarily fragmented and lacks a coherent sense of direction, purpose and unity."[19] Bob knew the topic: in 1968, he had written the Pulitzer Prize–winning series "Will Success Spoil the National Parks?" in the *Christian Science Monitor*. He was a member of the first White House Council on Environmental Quality (CEQ), an editor of *Audubon Magazine*, and one of the founding members of the NPS Unity Group with Mike Frome and myself.

In January 1993, at the beginning of the Clinton administration, the NPS responded to the Risser Committee's recommendations with a short report prepared for Associate Director Hester, Planning for the Future: A Strategic Plan for Improving the Natural Resource Program of the National Park Service. In fact this was merely a bland recitation of the Risser Committee's recommendations but put into bureaucratic form and noting that "the ability of the National Park Service to fully accomplish these resource preservation functions is not yet adequately developed."

In the early months of Roger Kennedy's tenure as NPS director, beginning in July 1993, I was appointed by DOI Secretary Bruce Babbitt as Roger's special assistant for policy and legislation, and thus enabled to push policy reform and compliance from *inside* the NPS for the first time. For the next eight years, one of my major tasks was working toward the Risser Committee recommendations. Roger also quickly brought in a super-smart NPS field manager, Maria Burks, to be the point person on his personal staff for implementing the recommendations of the *Vail Agenda*. Roger was new to the NPS and did not know many of the rising stars in the agency, but he did know Maria, who had been a key author of the Vail Agenda, and was working closely with his wife, Frances, on Civil War battlefield conservation.

Given responsibility for an aggressive legislative program to fulfill long-identified needs of the NPS, my staff and I took the unusual step of preparing a large package of proposed bills and amendments that we titled "The Spirit of Vail: A Legislative Program for the 103rd Congress." Included in this proposal was the long-recommended statutory mandate for science-driven decision-making in the NPS. With minor exceptions, none of these legislative proposals moved ahead in 1993–1994.

The main reason there was no action on the NPS science mandate proposal is that DOI Secretary Babbitt was seeking legislation to establish a new agency, the National Biological Survey, that would absorb all of the biological research functions of agencies in the department and from the NPS, the US Fish and Wildlife Service, and the US Geological Survey (USGS) in particular. At that time, the NPS had just over a hundred research-grade scientists. Virtually all were transferred to the new agency as a way to staff it, because opposition in Congress to this new agency was such that Babbitt was unlikely to get funding for entirely new research funds and positions. When the Republicans took over both houses of Congress after the 1994 elections, any hope of getting legislative approval for the new agency ended.

Instead, Secretary Babbitt established a new Biological Resources Division within the USGS by secretarial order, which he could do without congressional approval and that still resulted in transfer of NPS scientists to the USGS. What this reorganization failed to take into account was the difference between basic science research—the USGS mission—and applied research—the need of land management agencies. While both are important to expanding biological knowledge, the NPS needed readily available applied research, termed "usable knowledge" by NPS Chief Social Scientist Dr. Gary Machlis, to answer management questions and implement management decisions through its natural resource management specialists. For a time after this transfer of NPS science personnel, the NPS largely lost this critical scientific support capability. A few research scientists managed to remain within the NPS but had to shift into resource management.

Over the two years that the "Gingrich Revolution" controlled the legislative agenda in Washington, my time was largely taken with opposing their effort to establish a "park closing commission." The bill, HR 260, would have created a new commission appointed jointly by Congress and the administration to review the units of the national park system and recommend ones that ought to be deauthorized. This bill appeared to target small historical parks, large urban national recreation areas, and other sites that, in the opinion of Republican leaders, did not support a healthy tourism economy. Fortunately, this idea never gained enough support to be enacted, but it took a great deal of energy and time to defeat.

It was, therefore, not until the next Congress convened in early 1997 that a promising opportunity arose unexpectedly to seek a science mandate for the NPS. Senator Craig Thomas (R-WY) became the chair of the National Parks Subcommittee with Senator Dale Bumpers (D-AR) as the ranking member. Working very well together in a bipartisan manner, they decided that it was time to reform the long out-of-date NPS concessions law (see also chap. 6). What became Title IV, National Park Service Concessions Management Improvement, of the National Parks Omnibus Management Act of 1998 was by far the most visible and controversial legislation affecting the agency in that Congress. As it turned out, their bipartisan support for concessions reform allowed the resultant omnibus bill to also include other long-needed reforms, including a science mandate.

By that time, I was serving as senior advisor to the assistant secretary for Fish, Wildlife and Parks, Don Barry. Secretary Babbitt tasked me, as the lead negotiator for the administration, to work with the Senate committee on its multititle NPS management bill. Working out the amendments and compromises necessary to get concessions reform done took many deliberations, hearings, markups, and discussions. While that was underway, I worked closely and quietly with key Senate professional staffers Tom Williams, David Brooks, and Dan Naatz on the bill language that became Title II, National Park System Resource Inventory and Management. I drafted every word of this section of the bill, which did not need its own hearing, just the internal agreement of Senators Thomas and Bumpers. It was enacted without amendment or even comment when the final omnibus bill was passed and signed by President Clinton as Public Law 105-391.

Section 202 states "The Secretary is authorized and directed to assure that management of units of the National Park System is enhanced by the availability and utilization of a broad program of the highest quality science and information."

Section 203 states "The Secretary is authorized and directed to enter into cooperative agreements with colleges and universities . . . to establish cooperative study units to conduct multi-disciplinary research and develop integrated information products on the resources of the National Park System, or the larger region of which parks are a part."

Section 204 states "The Secretary shall undertake a program of inventory and monitoring of National Park System resources to establish baseline information and provide information on the long-term trends in the condition of National Park System resources."

Section 206 states "The Secretary shall take such measures as are necessary to assure the full and proper utilization of the results of scientific study for park management decisions."

Finally, the National Park Service had its statutory mandate for managing parks with science. By 1999, the NPS had responded to its new mandates with the "Natural Resource Challenge," a multiyear program of budget and staffing increases intended to carry out this mandate.

Unfortunately for this and other park management matters, George W. Bush was elected president in 2000. With his administration began another eight-year cycle of opposition to NPS preservation policies, ignoring the role of science in park management, neglect of NPS management needs, and budget and staffing cuts that provided an excuse for policy changes and opposition to any new parks.

Even before Bush was sworn in as president in January 2001, the incoming chair of the House Resources Committee, James Hansen (R-UT), wrote to president-elect Bush on December 27, 2000, with his long list of proposed actions and Clinton administration policies that he wanted changed. For the NPS, he expressed his opposition to a proposed ban on snowmobiles in Yellowstone and other parks, opposition to restriction on commercial air tours over parks, and opposition to a ban on jet skis on the waters of some parks. And he especially opposed the new edition of the NPS *Management Policies* manual, which asserted that resource preservation is the primary mission and requiring that recreation be limited to uses that are compatible with preservation.

During my last two years working for the NPS "from the inside," the Winter Use Plan for Yellowstone National Park became a major focus for my time. The political cross-hairs were set on the impacts of snowmobiles—both on wintering wildlife and the health of park employees—versus impact to the economic "engine" provided by the park's winter operations to several towns at the park's entrances, especially West Yellowstone, Montana, and Cody, Wyoming.

Winter in Yellowstone is a magical time for visitors. Steam rises much more visibly from the geysers and other thermal features for which Yellowstone is famous, and snow-encrusted bison congregate closer to the roads than in summer. The combination of this attractive experience and the increased availability of rental snowmobiles in the park's gateway communities created a major economic engine for winter use.

Yellowstone superintendent Mike Finley and his science and resource management staff had complied ample research data on toxic air pollution stemming from the two-stroke engines (also used in home lawnmowers) that were the industry standard for snowmobiles. The fuel of two-stroke engines is a mixture of oil and gas, and the engines burn it inefficiently, especially in winter. A toxic fog, snowmobile exhaust is very unhealthy to breathe, and the park entrance station rangers who collect fees and pass out information to visitors were exposed to it for hours at a time.

In addition, while snowmobiles are not allowed off of roads in the national parks (with rare exceptions),[20] in the winter Yellowstone's paved roads are packed down for snowmobile use by NPS maintenance crews. An unintended consequence of this grooming is that the bison love to make use of this much easier, compacted travel path and thus too often collide with passing snowmobiles—encounters known locally as "bison ping-pong."

During 1999 and 2000, NPS professional staff developed a full environmental impact statement (EIS) with alternatives, including one that would ban snowmobiles. The park held numerous public meetings, including in each of the three surrounding states, which I participated in as the DOI representative. Governors and congressional delegations of these states were nearly universally opposed to a snowmobile ban. Nevertheless, when the final EIS was released in 1999, the NPS "preferred alternative was to ban snowmobiles for individual usage, and instead offer concession-operated 'snow-coaches'" that would both carry multiple visitors into the geyser basin and be driven by trained operators who knew, through experience, how to avoid collision and other stressful conflicts with the park's bison.[21]

Completing the EIS process, without shortcuts or failure to fully comply with regulations, became a race against the end of the Clinton administration on January 20, 2000. The final EIS was completed and released in December 1999, after which the public had a thirty-day period for another round of comments before the Record of Decision (ROD) could be approved and signed. During the last week prior to the inauguration of President George W. Bush, there was a hectic scramble debating numerous wording changes to the ROD that passed between myself, Deputy Assistant Secretary Stephen Saunders, Assistant Secretary Don Barry, NPS Director Bob Stanton, Intermountain Regional Director John Cook, and Yellowstone Superintendent Mike Finley.

That last night, Friday, January 19, was the only time during my government service that I have pulled an all-nighter on the job, with some combination of the six of us passing final drafts back and forth via phone calls, fax machines, and rapid trips down the long hallways of the main DOI building through the night and early morning hours. In the end, early in the morning of January 20, the final text of the ROD was agreed to, and the signed ROD was hand-carried to the Federal Register Office for official publication before noon.

Ironically, multiple lawsuits were quickly filed against the final Winter Use Plan, led by former DOI Assistant Secretary for Fish, Wildlife and Parks Bill Horn, now in private practice. Among the first political appointees named to the DOI by the Bush administration was Paul Hoffman, previously executive director of the Cody, Wyoming, Chamber of Commerce, who had led the local opposition to the Yellowstone Winter Use Plan's snowmobile ban. Paul was to be the new deputy assistant secretary for Fish, Wildlife and Parks, a position from which he not only led reversal of the snowmobile ban but sought fundamental reversal of many NPS policies (see chap. 3) that had put conservation of park resources above visitor use whenever there was conflict. Horn and Hoffman both believed that visitor enjoyment, by whatever means, was the imperative that had to be met by the NPS.

The Yellowstone Winter Use Plan was quickly reversed by new DOI Secretary Gale Norton in 2001, allowing snowmobiles to continue to be used in Yellowstone and other parks. Norton, a protégé of former DOI Secretary Watt from the Mountain State Legal Foundation, had learned from him well and quickly made it clear that she would use the budget to change policy.

In December 2002, the Democratic side of the House Resources Committee, under ranking member Nick Rahall, published its analysis of how Secretary Norton was ignoring or denigrating science. Entitled "Weird Science— The Interior Department's Manipulation of Science for Political Purposes," it noted, "over the past two years, the administration has ignored, manipulated, challenged, suppressed, and dictated scientific analysis in order to implement an agenda harmful to the environment and to roll-back Clinton-era protections."[22] Among the many examples of ignoring science it cited, reversing the ban on snowmobiles in Yellowstone National Park was called out. While the Environmental Protection Agency was developing new pollution standards for snowmobiles, "a government whistleblower provided Congress with documents showing that in a draft letter to EPA, the Interior Department removed its own scientists' comments recommending snowmobile emissions be curbed to reduce haze and smog."[23] Secretary Norton reversed the ban on snowmobiles in Yellowstone and in fact allowed a 35 percent increase in daily entries into the park.

I was hired by the Rockefeller Family Fund's Environmental Integrity Project to help organize resistance to these policy manipulations at the NPS. In particular, we prepared a series of fact sheets describing each adverse park issue and recruited several recent career NPS retirees—Rick Smith, Bill Wade, and Mike Finley (who retired from Yellowstone in 2001), all former superintendents, among others—to organize professional opposition to these park impacts and speak with credibility based on decades of direct park management experience. "The Campaign to Protect America's Lands" was launched in May 2003 with a press conference and a letter to President Bush and Secretary Norton, signed by twenty-five retired NPS senior managers. Mike Finley, former superintendent not only of Yellowstone but also Assateague Island, Yosemite, and the Everglades, said, "the policies of President Bush and Interior Secretary Norton toward national parks are not based on science or sound conservation principles but purely on politics and favoring special interests."[24]

JONATHAN

The classic park ranger is an icon in America: tough, capable, friendly, knowledgeable, and there when you want to learn about the park, find the bathroom, or be rescued. I donned the green-and-gray uniform and classic ranger hat in 1976 and wore them proudly for forty years. Jokingly referred to as a "cradle-to-grave" ranger program, NPS training taught me to fight fire, carry a gun, make arrests, tranquilize rogue animals, splint a broken bone, greet visitors with a smile, pilot a four-wheel-drive, dynamite an avalanche, rappel down a cliff, climb mountains, ride a horse, raft a river, pilot a boat, scuba dive, and count elk droppings.

While all those ranger skills are really a lot of fun, and essential to making it through a typical day in a national park, my biology background allowed me to focus my time and energies on the resources of the park, on better understanding their status, and on how the NPS might be a better steward. In Destry's section of this chapter, he lays the foundation of the history of science in the NPS from the early days of George Melendez Wright, the scholarly advice of A. Starker Leopold, and the slow and often sporadic growth of science programs in the NPS. My story is from the inside of parks, and tells how science and resource management shaped many of the success stories in conservation over the last forty years.

For most of its history, up until the 1980s, the management of park resources fell to the rangers. As I pointed out above, rangers have a lot of skills and enormous responsibilities, and unless they have a background in science, the resources portion of their work was often subservient to the immediate:

a fire, a rescue, or visitors behaving badly. I witnessed this over and over as I rose through the ranger ranks, recognizing that, as parks became more complex and law enforcement required more professionalism, resources management was being left behind. Others both inside and outside saw this as well, and in 1980 Roland Wauer, chief of the Division of Natural Resources for the NPS, conceived of a new category of employee, the resource management specialist. Ro was able to convince the powers in Washington to invest in this idea, and he recruited and hired twenty existing employees from around the service to be the first class. I was selected from my district ranger job at Guadalupe Mountains National Park and was immediately transferred to Crater Lake National Park as a resource management specialist/trainee. For the next two years, our group of twenty gathered for graduate-level training in air and water quality, endangered species, vegetation management, NEPA compliance, and how to build a park science program. Our goal: to professionalize resource management with a core group of well-trained employees who were not rangers. Ro's goal was that some of us would rise to become superintendents (traditionally all rangers) and, maybe, one of us would become the director of the NPS. I frankly did not expect that person to be me.

In the winter of 1983, a small group from our trainee class traveled to Washington as part of our program. The goal was to meet the leadership of the Department of the Interior and the NPS, but we landed during one of the political purges overseen by Secretary James Watt. One of our first stops was the office of Division Chief Ro Wauer. It was empty, the government desk cleared, and the only adornment was an outdated government-issue wall calendar. Because of his relentless advocacy for the protection of park resources, Ro had been pressured to leave the Washington office (he went to the Great Smoky Mountains National Park as assistant superintendent) by Associate Director Richard Briceland, a move typical of the petty jealousies and party politics that are so common in Washington.[25] Word of his purge spread through the trainees and our resolve and commitment only deepened. We were then escorted to the office of Assistant Secretary for Fish, Wildlife and Parks Ray Arnett. As he stood with his back to us and suggested that we were no different than state game wardens, one of our group jumped up, turned to us, and opened his dress shirt to reveal a "Dump Watt" T-shirt underneath. Arnett never turned around and was never aware of the incident.

After that depressing tour of the offices in DC, our class of trainees gathered at the NPS Harpers Ferry Center to be addressed by Associate Director Briceland. The entire group was still fuming over the purge of our founder Ro Wauer, and our anger was focused on Briceland. In a rambling monologue, he began to explain how, in pondering the need for greater professional resource management in the parks, he had come up with the idea of the Natural

Resource Trainee program. He made no mention of Ro Wauer. Afterward, a group of us got together and considered resigning from the program in protest, but I convinced the group to stick it out. I said that if Briceland claimed the program as his own, even if we all knew that was a lie, he would support our completion of the training and the NPS would benefit in the long run. After sending Ro Wauer into exile, Briceland continued his purge of other park scientists, setting back park science for years.

Our class of resource management trainees graduated in 1984 and was followed by at least five more rounds of twenty or more trainees. Assigned and stationed in the field, these training programs stayed under the radar of changing political administrations and were able to survive. These well-trained and highly motivated employees spread out into the NPS, building programs at parks and slowly taking over resource management from the rangers.[26] Many rose to become park superintendents or regional and park chiefs of resources.

While participating in the Natural Resource Management Training program, Crater Lake National Park was my home base and the perfect example of this history of poor resource stewardship by a staff focused on the visitor. Crater Lake itself, an amazing body of water that filled a collapsed volcano, was the primary resource of the park and considered a world limnologic benchmark. Amazingly, the NPS itself had no program to monitor its condition. At best, there was a periodic testing of its clarity with a Secchi disk, a plastic circle the size of a dinner plate lowered into the lake until it disappeared, giving the observer a rough idea of water clarity. The rangers who conducted this test, without any study design, often laughed that the change in the ability to see the disk at various depths could either be a decline in the water quality or a decline in their eyesight!

Meanwhile, Doug Larson, a scientist employed by the Army Corps of Engineers, took his own samples and declared publicly that the lake was losing clarity. This caught the NPS flat-footed, and a war of words between the scientist and the park superintendent played out in the Oregon media. The NPS was defenseless as it had no data of its own to refute the allegations of the Army Corps scientist. As I learned through my career, there is nothing as effective as a media crisis to get funding. The Washington office granted funding to the park to launch its own research program. As the new park biologist, I had strong ideas on how to spend the funds but had to do battle internally to preserve them for the research program. In a classic case of the periodic downside of superintendent independence, Crater Lake's superintendent, Bob Benton, told me that it did not matter to him what I justified the funds for, once it came into the park budget, he could spend it on anything he wanted, including upgrading his office. With some help from the regional

Crater Lake National Park. Credit: Jonathan Jarvis.

director, I was able to convince the superintendent that a strong scientific program for the lake was the right thing to do—and would make him look good.

Although a biologist, I was still technically embedded in the ranger organization (the NPS had yet to organizationally recognize a park-based resource management division), and the rangers themselves were very supportive of my plan to build a science program for the lake. Their field skills, particularly those of Ranger Mark Forbes, were essential to building a research program on a lake that is 2,000 feet deep, sits on top of a cascade volcano at 7,000 feet above sea level, and averages 500 inches of snowfall per year. Enormous logistical and safety issues had to be overcome in order to do good science on a lake that was practically a mini-ocean.

Rather than just running a few years of research to refute (or confirm) the accusations of the independent scientist, I chose to create a program of monitoring and science that would continue into the foreseeable future. Key to that was engaging a research university, and Oregon State University was perfect. Dr. Gary Larson (no relation to Doug Larson), a renowned limnologist, became the principal investigator and designed a robust, ten-year program of research and monitoring. Dr. Larson suggested that we would never understand the biology of the lake unless we were able to sample the water in the winter, something that had never been attempted. With extensive planning, we were able to launch the very first winter expedition to Crater Lake to gather critical water samples when the lake became biologically active. Logistically, we had to fly by helicopter to be dropped off at Wizard Island, dig the boat-house (which we had built the previous summer) out from under twenty feet of snow, launch the research boat, pull deep-water sampling profiles from the thousands of feet of lake depth, preserve the samples, and hope to be rescued by helicopter when the weather was clear. We were stuck on the island for a few extra days and were prepared to assemble an inflatable boat, motor over to the 1000-foot wall of the caldera, and climb up an avalanche chute that is the only open path to the rim. But the weather did break and we were able to escape by helicopter with our water samples.

My mantra for Crater Lake National Park every morning was put up the American flag, open the gate, unlock the visitor center, and take the water sample—all part of the operation.

With the Crater Lake research program established with a complete water laboratory, research vessels, and a staff scientist, I moved on to North Cascades National Park as its first chief of Natural and Cultural Resources. One of my first actions was to invite a group of scientists from Oregon State University and the University of Washington to advise me on my priorities. One of those scientists recommended that I set up weather monitoring stations at multiple elevations through the park to collect data on climate change. This was 1986, when climate change research was just beginning. Unfortunately, I did not have the funding to start that kind of program.

I did, however, launch a variety of science programs at North Cascades but was again frustrated at the lack of basic knowledge of park resources. At the time, the National Park Service's planning operation at the Denver Service Center was conducting development of a general management plan (GMP) for the park, a document that would serve for at least twenty years as a guide to management and development. When I saw the base maps for the draft plan and noted that there were colored bubbles on the pages denoting what areas of the park could be developed for facilities and which areas would be protected, I looked for any resource data that might support such planning decisions. I found none. I suggested to the team that perhaps we should invest a little of that planning money to determine exactly what resources might be inside those bubbles. I was told that park planners did not need any scientific information about the resources in order to make decisions. I was accused of wanting to "count squirrels" and my ideas were dismissed. The poorly supported draft plan was released to the public, severely lacking in any resource data.

Almost immediately, the local environmental group, North Cascades Conservation Council (NC3) and the Sierra Club Legal Defense Fund (SCLDF) sued on the grounds that the plan and its environmental impact statement were devoid of any real resource data upon which to make decisions. When the

Justice Department attorneys reviewed the case, they told the superintendent that the NPS would lose the lawsuit and recommended that the NPS enter negotiations to settle the case. I was designated by the superintendent to lead the settlement negotiations. Dave Fluharty, president of NC3, and Stephan Volker,[27] an attorney from the SCLDF, were on one side of the table. I was on the other with a Justice Department lawyer, and we negotiated the details of a settlement over the period of several weeks. On the evening prior to each meeting, I would meet privately with Fluharty and we would lay out all the demands that NC3 would request of the NPS. I had my long list of basic science needs: wetland mapping, vegetation inventories, stream water quality, soils and geology, wildlife monitoring, and so on. The following day, during the negotiation, Fluharty would lay out these exact demands, the Justice attorney would turn to me and ask if I accepted these as new requirements, and I would (of course, since I created the list) say yes. The final settlement agreement and consent decree from the court had all the science that I had originally wanted the planning team to establish in the first place. This may have been a backhanded way to get baseline information for the park, but it worked. That investment significantly improved the park plan and became the basis of the park's robust resource management program. It also fundamentally changed the way national parks did their general management plans. The planning teams now recognized there should be a baseline of resource information for each park and supported funding for parks to build that database.

During North Cascades' GMP process, I received my first lesson on how some conservative politicians view science. Presenting our plan to map resources within the Lake Chelan National Recreation Area to the Chelan County Commissioners, one conservative member stated that aerial photography of land that included any private lands (and there were private inholdings in the recreation area) should be considered an invasion of privacy and illegal. I pointed to the large aerial photograph of the City of Chelan that hung in the conference room where we were meeting and said, So that aerial photograph should be illegal? He said yes.

It was during this process that I also learned about the fringe elements of our society who fundamentally disagree with the government managing national parks. At the public meeting for the park's general management plan in Wenatchee, I was serving as the timekeeper for each member of the public who came to the microphone. There was a large crowd and so we allocated about three minutes for each and, with less than one minute left, I would hold up a sign to finish up their comments. The first three speakers were representatives of a far-right group that adhered to an interpretation of the 1878 Posse Comitatus Act, where the only authority over the people is the elected sheriff. The first of the three explained that since the Chelan

County sheriff had not authorized this meeting, the gathering was unlaw-
ful, the plan illegal, and the very existence of the National Park Service was
against the law. The second speaker pointed out the plan made references
to the relationship and border with Canada as well as the designation of an
International Biosphere—and therefore the United Nations would soon be
running the park. The third speaker, clearly carrying a sidearm under his
shirt, called for the crowd to gather outside and hang the park superintendent, John Reynolds. I looked over at John and he was sweating. At that point,
I held up the sign that the speaker's time was up, but I held the sign to the
side, not over my chest, just in case he might decide to extend his speaking
time with use of his firearm.

With my career now focused on a professional track in resource management, I developed relationships with a number of scientists who were doing
quality research in the western national parks. Most of these individuals were
based in the regional office or at what were known as cooperative park study
units (CPSUs) at the Universities of Oregon, Idaho, Washington, and California. Gary Machlis, Don Field, Ed Starkey, Gerry Wright, Jim Agee, Gary Larson, Shirley Clark, Stephanie Toothman, and Jim Larson all became friends
and mentors, guiding me in my stewardship of park resources. It was from
this group also that the social science program of the NPS grew. Led by Don
Field and Gary Machlis, the NPS began to better understand visitor demo-
graphics, activities, and attitudes, and, supplied with that information, the
NPS became a better host and steward. The economics program was also
then established, giving the NPS strong economic data upon which to articulate positive impact to gateway communities.

Over beers, we often discussed these scientists' findings, their implications, and how management might shift based on their research. From them,
I learned both the opportunities and the limitations of science in park management. Most of them had learned to walk in both worlds and displayed em-
pathy when the park was faced with a complex, political resource decision
without much scientifically based information. It was a conundrum I faced
over and over, but I knew that I could have the most influence on the organization if I rose to the rank of park superintendent, a field still dominated
by rangers. I applied to every superintendent job I could find and, in 1991,
landed Craters of the Moon National Monument in Idaho.

Craters of the Moon National Monument sits atop the Snake River plain,
surrounded by BLM lands that were grazed by cattle, and nestled in Butte
County, where the population is roughly one person per square mile. The
local ranchers and business people are politically conservative, and the federal government is unpopular. But there is a deep love and appreciation of
Craters of the Moon and so I was welcomed into the community, our kids

into the schools, and myself into the local Rotary. Our weekly Rotary lunch of chicken fried steak and weak coffee gave me great insights into rural Idaho, local politics, and a community that felt abandoned by the government. As a first-time superintendent, I got to experience the autonomy that comes with the job and therefore created a new division of resource management and hired its first chief, Vicki Snitzler. We worked on a range of issues including restoration of native species, monitoring bat populations, abandoned mine restoration, and water rights adjudication. I also laid the foundation for expansion of the park to include the entire lava flow on the Snake River Plain by including a map of the resource in the park's management plan. That map became the basis of major expansion by President Clinton in 2000 and by Congress in 2002.

After three years at Craters of the Moon, I moved to Alaska to be the superintendent of Wrangell-St. Elias National Park and Preserve (see chap. 2). For the first time in my career, I had a real, practicing scientist on my staff, Dr. Kurt Jenkins. Kurt was studying the predator-prey relationships of caribou, grizzly bears, and wolves. We often discussed how to balance the "research question" that would gain new knowledge from his work and the "management question" that would give me the information needed to make decisions about harvest limits on caribou for subsistence by the local Ahtna people. The park resource team, led by the highly capable Russell Galipeau, always seemed to find that sweet spot that achieved both.

But, then, as Destry described, Interior Secretary Bruce Babbitt launched what I believed to be an ill-conceived reorganization that rounded up all the bureau scientists and attempted to create a new DOI agency, the National Biological Survey. Babbitt's "big mistake," as those of us in the NPS called it, was a political disaster. Conservatives in Congress interpreted the title "Biological Survey" as a plan to survey private lands for endangered and sensitive species and take their property. Congress refused to create the new agency, forcing Secretary Babbitt to roll our former NPS scientists into the USGS. Dr. Jenkins was removed from my park staff and moved to Anchorage along with his salary and support funding from the park budget. Due to the objections to government scientists doing surveys that might impede development, Dr. Jenkin's funding was immediately slashed, severely restricting his ability to travel to the park to do research. The NPS science program was set back at least a decade as the direct connections of scientists to managers became more difficult. While USGS had plenty of scientists, some of whom did research in national parks, their understanding of how that science would apply to decision making was lacking.

This reframing of natural resources management was crystallized in Richard Sellars's seminal book, *Preserving Nature in the National Parks*.[28] The book had a profound effect on the leadership of the NPS at the time, even point-ing out an internal cultural resistance to prioritizing science and preserva-tion over visitor use. Recognizing that change was necessary, Director Bob Stanton launched the Natural Resource Challenge in 1999.[29] The Challenge resulted in an increase in the NPS budget of nearly $80 million dollars. Over 200 national parks that had a significant set of natural resources now had both staff and funding to carry out a professional program. Over the course of several years, resource inventories were conducted and specific indicator species selected for long-term monitoring, setting in motion the accumula-tion of trend data that would later reveal significant impacts, including those from climate change.

Fortunately, the embedding of the natural resource management staff had taken deep roots and there was no going back. At some point, we took off the metaphorical protest armbands over losing our scientists and began to rebuild from within. Still administratively prohibited from hiring scientists subject to "research grade evaluation," the NPS was not prohibited from hir-ing PhDs to run its resource programs.

The Natural Resource Challenge also launched the Cooperative Ecosystem Study Units (CESU), a successor to the Cooperative Park Study Units men-tioned earlier. The CESU network, based at major universities, brought to-gether not just the NPS but other federal and state land and resource agen-cies in a framework for cooperation at the ecosystem scale.

As the superintendent of Mount Rainier National Park from 1999 to 2002, melting glaciers provided an opportunity to begin to talk to the public about climate change. I also engaged the community of geologists who study active volcanoes, bringing their expertise to the safety and evacuation preparations needed in the valleys, homes, and communities that surround Mount Rainier. Far from extinct, Rainier is the most dangerous volcano in North America.

In 2002, I was selected to be the regional director of the Pacific West Re-gion. From my office in Oakland, California, I now had oversight of fifty-eight park units, from Yosemite to Haleakalā, from Golden Gate NRA to Manzanar, and spread over Washington, Oregon, Idaho, Nevada, Hawaii, Guam, Saipan, and American Samoa. Now physically and geographically separated from the resources of the parks, I needed to have a voice of science in my ear, unfiltered by the political and operational noise that dominates the role of a regional director. I asked Dr. Dave Graber to be my science advisor. A former student of A. Starker Leopold, field hardened and curmudgeonly, Graber was perfect to make sure I stayed on the right course. I could call him at any time, and he could call me, and I could dispatch him to investigate the status of critical issues like the plight of the Devils Hole pupfish of Death Valley, bear man-agement in Yosemite, or the island foxes on Channel Islands.

With Dr. Graber's able assistance, we began the first series of workshops in the National Park Service to address the impacts, role, and response of the agency to global warming and subsequent climate change. These workshops were taking place during the administration of President George W. Bush and Secretary Gale Norton, both of whom were openly antagonistic to the science of climate change. In Washington, climate scientists such as James Hanson were being grilled over the hot coals of congressional hearings. We were fortunate to be on the West Coast, 3,000 miles away. Our workshops revealed that parks were already experiencing climate change effects: Cascade mountain snow packs were melting from warm autumn rains, in the Sierras the snowpack was drier and thinner, we had longer fire seasons and hotter fires, beetle-kill of forests was expanding, and coral was bleaching on the coasts, to name a few. These workshops formulated my view that the NPS had a very clear role in addressing climate change through monitoring, mitigation, adaptation, and especially public education about how parks were being impacted by climate change.

By this time in the agency's history, science and its application to park management were becoming established. This was in part fueled by the budgetary and program development of the Natural Resource Challenge of 1999, with its increased budget and launch of robust inventory and monitoring programs. Excellent science was applied to reorganizing the facilities in Yosemite Valley in recognition of periodic flooding and scary rock falls from the granite walls. Good science called for the return of natural fire to the Sierra forests, rebuilding their resiliency, biodiversity, and resistance to the impacts of catastrophic wildfire. Good science showed the path to the restoration of the California condor, the captive breeding of the endangered Channel Island Fox, and the need to create marine protected areas that were "no-take" reserves free from both commercial and sport fishing. Good science did not "make the decision" in these cases but informed the decision with what Dr. Gary Machlis called "usable knowledge." With the traditional rangers now retiring early due to their twenty-year eligibility, fewer applied for the park superintendent positions. I began to hire resource managers, historians, archeologists, and biologists to lead the parks as an investment in their resource stewardship. In my seven years as regional director, for the fifty-eight parks I supervised, I hired fifty-seven new park superintendents. Only a few were former rangers.

But just because we were generating good science and attempting to apply it did not mean we were free from controversial decisions. One could argue that the science made the issues more controversial, and, as I quipped to a political appointee, "knowledge limits your discretion." In other words, it is easier to justify a decision when there is ignorance of the consequences. For the forces that look to parks as only economic assets or resources to be exploited, science that indicated resource damage was not welcomed warmly. This was particularly true of political appointees in Republican administrations. Deputy Assistant Secretary of the Interior Paul Hoffman was a good example of someone using his position to suppress science that did not agree with his ideology. At Mojave National Preserve, where native species have evolved to survive in a harsh desert environment, Hoffman wanted the NPS to install artificial water holes, known as guzzlers, throughout the park for the benefit of game animals such as deer. This was in direct contradiction of the science that showed guzzlers can increase predation on the endangered desert tortoise by ravens. When Mary Martin, the park superintendent, refused, Hoffman demanded that she be removed. When I refused to remove her, Hoffman engaged local Congressman Duncan Hunter to hold field hearings where the park staff and I were berated before an audience he had handpicked. Congressman Hunter loudly accused me of violating the California Desert Protection Act that had established Mojave National Preserve by not allowing the state to install guzzlers. He was unaware that I had served on the team that wrote that legislation and that the specific portion he referenced actually pertained only to the BLM and not the NPS. When I pointed this out to him, he only became even more irate.

Another political approach was to attack the scientists. The oyster farm at Point Reyes National Seashore provides an excellent example of this challenge. Prior to the establishment of the national seashore in 1972, a commercial oyster farm operated in the estuary known as Drakes Estero. Much of the lands that were incorporated into the seashore were dairy farms and, when purchased by the federal government, the NPS granted leases to the owners to continue to operate, with the recognition that the farms contribute to the local agriculture community and were generally compatible with the history and stewardship of the seashore. Not so with the Johnson Oyster Company, which the NPS purchased and granted only a term permit with an end date with the full intent to remove it and restore the estuary to its natural state. This story is fairly well told in *The Oyster War* by Summer Brennan.[30] When in 2005 the historic owner sold the remaining five years of the permit to a local rancher, who announced his intent to keep the oyster operation going indefinitely, the NPS launched into a battle that involved science, politics, local activists on both sides, the Inspector General, the National Academy of Science, and the State of California agriculture agencies.

When the park staff attempted, rather clumsily, to prove that science demonstrated significant impact to the resources of the estuary, the science itself and the scientists who conducted the research as well as the park managers and myself as the regional director all became targets of attack. All of us were accused of and investigated for scientific misconduct, with the oyster farm supporters accusing us of intentionally misrepresenting the science. The peer-reviewed science of impact from the oyster farming on the resident harbor seals and the extent of the sea grasses came under direct attack by both a paid lobbyist and a local scientist with no qualifications in marine ecology. Under pressure from Senator Dianne Feinstein, NPS Director Mary Bomar called me weekly to ask that I remove the park superintendent, Don Neubacher. I refused every time and defended Don and his team of scientists. As a compromise to Director Bomar and Senator Feinstein, I commissioned the National Academy of Science to assemble a team and evaluate the science behind the oyster farm and its potential impacts to the resources of the seashore. The report, when eventually released, concluded that the oyster farm could be compatible with the biological activity of the estuary, but it ignored the standards set by the wilderness designation within a unit of the national park system. Staff at the academy revealed to me that the report was an "embarrassment" and one of the worst they had ever issued.

In 2009, I was nominated to be the director of the National Park Service by President Obama, and the oyster lobbyists ramped up their opposition to my confirmation. They launched repeated filings with the Inspector General, accusing me of scientific misconduct among other things. The investigations were carried out by the Department of the Interior, the White House, and the Senate and found no evidence of anything except a smear campaign by the oyster industry lobbyist.

Ultimately, after an exhausting battle over the science, the decision on the future of the oyster farm boiled down to one of law and policy: Would Secretary of the Interior Ken Salazar decide in favor of the National Park Service and support the removal of the farm, or decide in favor of the oyster farmer? There was enormous pressure to support the oyster farmer from members of Congress, including Senator Feinstein, who chaired the Interior Appropriations Committee in the Senate. Feinstein attempted to legislate the oyster farm's continued existence but was backed down by Senator Jeff Bingaman (NM). I had invested a great deal of my reputation for good science and good policy in this case and was prepared to resign as director if the decision went the wrong way.[31] Secretary Salazar sought input from his senior staff and made the decision to remove the oyster farm. His decision was challenged all the way to the Supreme Court, which remanded it back to a lower court that upheld the NPS position for removal but told all parties to settle. The oyster farmer agreed to walk away, leaving millions of pounds of oyster shells, racks, and debris behind for the NPS to clean up. We accepted that decision, spent millions of dollars on the cleanup, and today the estuary is clean and thriving without the noise of boats or the impact of miles of oyster racks.

There were lessons here too, about the limitations and vulnerability of science in a hot political debate. There were scientists on both sides and unfortunately the debate at times turned very ugly. In addition, science by its very nature is a competition of theories, strengthened by debate, but in the arena of politics, the debate can be used to undermine good science that backs up a decision for the conservation of a park resource. As a result, it is important to note that science informs the decision but it often ultimately comes down to policy and law.

When confirmed as director, I called upon Dr. Gary Machlis to serve as my science advisor, the first ever such position. A trusted friend, I knew Gary would be unafraid to call into question my decisions or to sense when a proponent, even inside the NPS, was making a pitch that was not supported by science.

Having been accused of scientific misconduct during the oyster wars and investigated by the agents of the Inspector General's office, I knew the NPS needed a new approach. Well-meaning and honest, the law enforcement agents in the Inspector General's office were unqualified to evaluate the nuances of scientific misconduct or the integrity of a scientific report. Also, recognizing that during the last Republican administration, political staff rewrote biological opinions of the Fish and Wildlife Service, the first task I assigned Gary was to develop a robust scientific integrity policy for the Department of the Interior. In a stroke of brilliance, Gary wrote the policy so that it applied not only to scientists but to their (nonscience) supervisors as well. Now when an accusation is made of potential misconduct, the investigation is done by scientists who are the designated scientific integrity officers. The policy was adopted and was the first of its kind in the Obama administration.

Just as we were getting our agenda started, the Macondo Well in the Gulf of Mexico blew its stack, resulting in the Deepwater Horizon oil spill. I was dispatched to serve as an incident commander in the interagency command in Mobile, Alabama. We were responsible for cleaning up the ocean and beaches over a 200-mile stretch across the coasts of Florida, Alabama, Mississippi, and Louisiana. Unlike most disasters that are episodic, such as a hurricane or flood, this one continued every day for five months. Every day was Groundhog Day. The scientists in the incident command focused on the weather and the chemistry of the oil but not so much on the resources at risk. Dr. Machlis suggested we bring in a team of new scientists, all of whom have a history of research in the Gulf. This was the pioneering of a new concept, strategic science during a crisis, where science could drive our actions in the aftermath. Gary's Strategic Science Group (SSG) predicted a wide range of cascading effects from the oil spill, and their report created a framework for restoration of the Gulf. In the middle of this crisis, I was riding in a car with

Secretary of the Interior Ken Salazar, on our way to Camp David for a meeting about the spill. Salazar was on the phone with the president of British Petroleum (BP) and demanding 1 billion dollars, up front, to start the restoration process. He got it. The SSG work then guided much of the investment of those billions paid out by BP.

Management of the 400-plus units of the National Park Service is fraught with complex issues (what I called "hardy perennials") that just seem to never go away but spring back year after year. Examples include the restoration of Everglades National Park, where a century of attempts to drain it needs to be reversed. Recognition of the ecological importance and the simultaneous collapse of the Everglades started in the early 1970s. Political action began with the State of Florida and the Department of the Interior and eventually resulted in the Comprehensive Everglades Restoration Plan in 2000. At the core of the restoration is science, applied to the quality, quantity, and timing of the freshwater flows through what is known as the "river of grass." This underpinning of science for restoration also developed in other national parks. And the approach came under periodic political fire, such as when President Bush's Secretary Gale Norton attempted to close the Everglades Restoration office.

On my first trip down there as director, I set up a meeting with *only* the scientists to hear their perspective on the restoration of the Everglades. For the previous eight years, the restoration had been in the hands of the Army Corps of Engineers, and, while they ignored the science, they sure liked to build things like pump stations. The major problem with a pump station is that it costs a lot of money to build but the decision to actually turn it on is political. More often than not, after one is built and the Corps hosts a ribbon-cutting for the politicians who got them the money, the pump station is mothballed and sits idly for years. When I asked the scientists who should lead the restoration effort under the Obama administration, they responded "no one from the Corps, no one who has ever worked for the Corps or has anyone in their family that works for the Corps"!

From my perspective science had to be at the center of solving these complex issues. Snow machines in Yellowstone, euphemistically known as "Winter Use," had plagued the last six park superintendents, and now a viable solution was on the table, based on the science of sound and its impact to wildlife. The controversies surrounding grizzly bears in Yellowstone and Grand Teton go back to the days of pioneering bear biologists John and Frank Craighead, who made policy recommendations in the late 1950s. Those controversies continue to this day. The park and the state agencies were in a science duel over how to estimate the bear population, with the potential that imaginary bears could be created with math and real bears shot for sport. Other complex issues include Colorado River flows through the Grand Canyon, where science would tell us that periodically the flood gates of the Glen Canyon dam should be opened and allow the river to roar as it did in the past. With science as our guide, Secretary Salazar opened those gates and restored miles of beaches and habitat along the river. Science told us that we needed more natural fires in the Sierra parks, that wolves were genetically narrowed at Isle Royale and might soon wink out, and that historic fish stocking of high mountain lakes impacted native invertebrates.

These are highly charged policy, political, economic, legal, and resource battles that rise and fall each year, challenging the leadership of the NPS. Framing each one through a lens of science did not necessarily solve them, but it provided insight, political defense, and a rationale for the decision.

From 1982, when I transferred to Crater Lake National Park, until 2009, when I moved to Washington, DC, as the director, I had been working in or supervising national parks in the mountains, deserts, and forests of the Pacific West. With the eye of a natural resources specialist and an ear tuned to the advice and observations of scientists, I knew that climate change was real, it was happening, and it was having a profound impact on the national parks. For instance, for most of recorded history, winter snow accumulated in the Cascade Mountains, and, when the spring rains came, the snowpack soaked up that rain like an enormous sponge and then slowly released it, feeding the streams that provided essential water to fish, wildlife, homes, cities, and agriculture in the valleys. But starting around the mid-1990s, we began to witness a shift, with rain coming in the fall onto a thin snowpack. Instead of absorbing the rain, the snow melted and contributed to the volume, and enormous floods ripped down the river valleys. At Mount Rainier, such an event wiped out the Sunshine Point campground, scouring to bedrock a popular recreation site that had stood for over a hundred years.

Climate change was a new kind of existential threat to the national parks. The policy paradigm that had persisted for fifty years, generated from the Leopold Report, was one of putting all the pieces of nature back in place and then allowing natural processes to drive the ecosystem. This was not a hands-off approach, but it did rely on the basic principle that natural systems would provide some relative balance. The NPS would defend the resources from external impacts with whatever tools it had, supported by advocacy organizations and the courts, and then assume the integrity of the park resources was maintained. But with climate change, created by the burning of fossil fuels such as coal, oil, and gasoline, pumping carbon dioxide into the atmosphere and warming the planet through a "greenhouse" effect, the agency's ability to confront the impact directly was very limited if not impossible. The NPS needed a new paradigm and policy so that parks could address climate change through education, mitigation, monitoring, and adaptation.

As detailed in chapter 3, in June of 2011, I called Dr. Machlis to my office and asked him to revisit the Leopold Report, to update it to meet this new challenge. Dr. Machlis immediately recognized that this would be like rewriting any sacred text that was ingrained into a culture. Dr. Machlis and I engaged Associate Director for Cultural Resources Stephanie Toothman to ensure that the revisit of the Leopold Report would also address the stewardship of cultural resources and include the recognition of traditional ecological knowledge of native people. An integration of the natural and cultural resources was essential to this new approach.

The Revisiting Leopold Report recommendations were converted to NPS policy in Director's Order 100, which included a recognition of Native American traditional ecological knowledge alongside Western science. In addition to establishing a framework for adapting to climate change, it also required that all park superintendents demonstrate "scientific literacy" commensurate with their responsibilities, including those at cultural and historical parks:

> To further the stewardship goal in management decisions, the NPS will require superintendents and those who aspire to leadership positions to possess scientific literacy appropriate to their positions and resource management decision-making responsibilities.
>
> *Scientific literacy* is the knowledge and understanding of scientific concepts and processes, an understanding of the strengths and limitations of scientific findings, and the appropriate application of scientific research to management and policy issues.[32]

Director's Order 100, placing science at the core of our stewardship, was rescinded by the Trump administration on August 16, 2017, portending a new era of political attack on science, scientists, and park managers who use science in their decisions.

In the four years of the Trump administration, the rollback of science was unprecedented, comprehensive, and not limited to just the Department of the Interior and the National Park Service. References to the impacts of climate change were removed from park management plans, websites, and interpretive displays. Park scientists were prevented from speaking to the media. In a well-documented case, Dr. Maria Caffrey was pressured to rewrite her scientific report on impacts from sea level rise. She testified before Congress on her treatment.[33] In another, NPS climate scientist Dr. Patrick Gonzalez, a member of the Nobel Prize–winning International Panel on Climate Change, was pressured to remove references to human causes of global warming in his research publications. In both cases the scientists refused, but Caffrey resigned and Gonzalez was formally disciplined by his supervisors for speaking to Congress and the press, in spite of the fact that he did so in his role as professor at the University of California, Berkeley. In a detailed report of the many rollbacks of environmental protections carried out by the Trump administration, Dr. Gary Machlis and I revealed an overarching attack on science far beyond just the NPS.[34] Examples include the restriction of science in the promulgation of regulations to protect human health to only those that make the raw data publicly available, eliminating centers for research on children's health, and ignoring federal scientists' recommendations on clean air, clean water, and endangered species.

Without Usable Knowledge from Science, Park Management Is Left to the Uninformed

Our two narratives paint a rather negative history of fits and starts for science in the National Parks. Setbacks and impediments have resulted from untimely deaths, political whims, and internal culture. Well-meaning reorganizations of the department that attempt to "balance" the portfolios of the various bureaus usually result in a loss to the National Park Service, which is viewed often as a larder of resources for the leaner agencies. At its worst, ideologically conservative appointees attack science in what has been characterized as willful ignorance, editing documents and rewriting peer-reviewed reports. If they can't stop the science, they turn on the scientists and transfer or remove them. The result is that talented young scientists like Maria Caffrey and Patrick Gonzalez question their career choices and look to move on, creating a brain drain.

We would be remiss to not recognize, in spite of all the setbacks, that the NPS today has a robust inventory and monitoring program, many professional scientists hidden in the ranks, and dozens of restoration projects underway. But the heart of the setbacks always goes back to the political shift the NPS experiences with the change in administration, when the conservatives take over and view our national parks as nothing more than economic engines with little other intrinsic value. The new paradigm of using science to guide park management in light of climate change is even more politically imperiled. Parks resources are changing, moving, and hardly static. This opens up the parks to political suggestions for more aggressive manipulation, such as the logging of trees that have died from warmer, drier winters.

For the agency to truly meet its mandate of "unimpaired for the enjoyment of future generations," built upon the best available science, the National Park Service must be freed from the political whipsaw and given greater independence.

FIVE

ECOSYSTEM THINKING REQUIRES COLLABORATION

No national park—not even Wrangell-St. Elias—is big enough to be a self-sustaining ecosystem. Even though the national parks of Alaska are huge, salmon that spawn in the park's rivers migrate into the Pacific Ocean for much of their lives. Caribou herds roam from the parks onto other state and federal public lands and even into and out of Canada, and migratory birds travel to the parks for summer breeding but head south when winter arrives. All national parks are part of much larger ecosystems populated by farms, cities and towns, industrial areas, working forests, highways and utility corridors, and private homes. Conservation of the national parks is inextricably linked to the health and sustainability of the ecosystems of which they are just a part.

The key to working outside the park boundaries is through partnerships with other agencies, not-for-profit organizations, for-profit businesses, gateway communities, and philanthropic organizations. In other words, the National Park Service cannot do it alone. The agency must be agile and adept at working with a wide array of organizations at the local, regional, and national level that support its preservation mission. Over our ninety-plus years of combined experience, we have worked with many organizations, other federal agencies, and Friends groups as well as the collective national park system. As if that were not complex enough, the changing political dynamic every four years creates chaos with partners: in some periods they are treated as equals, in others as contractors, and at the worst times as enemies.

A highly desirable, new direction for the NPS is to undertake joint management agreements with culturally affiliated Native American tribes. Traditional indigenous people already think in terms of ecosystems, and the cultural connection to their ancestral landscapes is inextricably tied to their very being as a people. The NPS ought to be committed to better engaging them in a shared responsibility to conserve the parks unimpaired and to building widespread recognition in the general public that both the national parks and these indigenous peoples will be here in perpetuity.

In this chapter, we will describe the history of these relationships and why independence for the NPS would create new, more robust partnerships as well as more sustainability at the landscape scale, for perpetuity's sake.

DESTRY

The first park partners were the so-called nonprofit cooperating associations, nongovernmental organizations (NGOs) chartered early in the twentieth century by the NPS to publish park-specific educational materials and operate in-park bookstores. Under their operating agreements, these groups donate their profits back to the NPS to support primarily research and education programming. There are some park-specific cooperating associations, and there are conglomerate cooperating associations that serve many parks, such as Eastern National. Most have been successful, donating hundreds of thousands of dollars to the NPS annually.

The second group of NGO partners arose in the 1980s during the severe budget cuts begun by the Reagan administration and DOI Secretary James Watt. These new partners came to be known variously as Friends, Funds, Conservancies, or Foundations and were generally associated with one national park unit. Each organization entered into a fund-raising agreement with the NPS to raise private philanthropic funds to support specific park projects and programs. Some of these, such as Friends of Acadia or the Conservancy for Cuyahoga Valley National Park, have raised millions for NPS projects, ranging from land acquisition to construction and restoration. A common purpose of partner funding is to support a conservation service corps of young men and women working to supplement NPS employees on projects like deferred maintenance, storm damage restoration, resource management projects, or visitor education support, facilitated through a formal cooperative agreement with the agency.

Some of this wave of park partners has evolved further since the 1990s. Typically a park's cooperating association merges with its Friends group, creating organizations like the Golden Gate Conservancy and Yosemite Conservancy, for example. These conservancies have expanded the scope and scale of what park partners do to include direct construction project management and direct operation of some visitor programming. They have dramatically scaled up private funding support for national parks.

In 1967, the US Congress established the National Park Foundation (NPF) to serve as the national philanthropic partner to the entire NPS. Congress authorized an NPF board of directors, all appointed by the president of the United States with the Secretary of the Interior as the chair of the board and the director of the National Park Service as the board secretary. This early NPF model proved to be problematic as the board was viewed as political, with positions handed out as plums to supporters of the president. Interest in and support for the National Park Foundation has also waxed and waned with the interest of the Secretary of the Interior, who generally delegated this responsibility to a junior political appointee but regularly handpicked the foundation's executive director. The result was, according to one outgoing board member in 2009, that the National Park Foundation was raising about as much philanthropic funding "as a small-town historical society."

However, a small measure of independence for the NPF was finally achieved in 2016 as part of the NPS Centennial Act. The NPF authorization was amended to make the DOI Secretary and NPS director nonvoting board members, and it included a direct appropriation for the NPF for the first time.[1] As the new CEO beginning in 2015, Will Shafroth and his board revolutionized the NPF with their singular focus on the 2016 NPS Centennial. In close collaboration with Jonathan as NPS director, the NPF raised an unprecedented $500 million in philanthropic financial support for the NPS mission.

A third category of partnerships includes public-private collaborations in which the NPS is only one component of the management of a larger landscape and is entirely dependent upon other partners for success. There are a few NPS-NGO conservation management partnerships that have been successfully authorized by Congress for some unusual or complex places, including the Appalachian National Scenic Trail, the Tallgrass Prairie National Preserve in Kansas, and the Shenandoah Valley Civil War Battlefields National Historic District partnered with Cedar Creek and Belle Grove National Historical Park.

I have had a long and extensive involvement with the Appalachian National Scenic Trail (AT) covering most of my life. The AT mountaintop route passed within sight of the town where Jonathan and I grew up. My first twenty-mile hike as a teenager was on a section of the AT in the Jefferson National Forest that, as an adult, I successfully lobbied Congress to have designated as the James River Face Wilderness. While working at the NPCA in the late

Appalachian Trail hiker in Shenandoah National Park. Credit: Destry Jarvis.

1970s, DOI Secretary Andrus appointed me to the Appalachian National Scenic Trail Advisory Committee for a four-year term. After leaving DOI/NPS in the early 2000s, I was elected to the board of the Appalachian Trail Conservancy (ATC) where I served six years, including as the vice chair. Over all these years I have hiked sections of the AT in each of the fourteen states through which it is routed, though I have not yet done a full five-month "thru hike," as it is known, of the entire 2,190 miles.

The best existing NPS model of effective comanagement is this one between the NPS and the Appalachian Trail Conservancy (ATC). The Appalachian National Scenic Trail, a continuous footpath (only—no bikes, horses, or motors) along the crest of the eastern mountain ridges through fourteen states from Georgia to Maine, was envisioned, built, and maintained today by private citizens, primarily serving as volunteers. In 1984, the National Park Service and the ATC entered into a cooperative management agreement whereby the ATC assumes direct management responsibility for day-to-day trail maintenance and resource management, education, and interpretation, while the NPS carries out law enforcement, planning, regulations, most AT-specific land acquisition, and provides management funding to ATC, which is complemented by private donations and grants.

While the narrow AT corridor is about 99 percent protected, mostly by federal and state land acquisitions, there is much more to the quality of the AT experience than can be preserved by this narrow corridor, including its expansive vistas, ecological integrity, wildlife abundance, and the preservation of thousands of historic properties along the route. The NPS/ATC partnership is by far the most extensive ever utilized by any federal land agency for day-to-day management of public lands.

This AT model has applicability to numerous other sites under NPS management but has not been undertaken elsewhere to any significant extent. Despite a few examples, the model of direct NGO comanagement is relatively untried, though it represents a viable approach that ought to be applied in other areas where there is a strong and well-established NGO partner organization.

The fourth category of partnership is more common but still not fully formed. In these, the National Park Service must forge partnerships with states or local governments to manage the natural and/or cultural resources of areas within or contiguous with park boundaries but not under NPS control. Many of the national seashores and lakeshores, such as Assateague, Fire Island, and Padre Island, or Indiana Dunes (now a national park) have contiguous state parks, with which regular comanagement activities occur. Redwood National and State Parks have taken this even further through a formal comanagement agreement, with shared offices, maintenance yards, interpretive programs, and resource management functions. This is a great model for application in numerous other NPS sites.

Beginning with the designation of Golden Gate and Gateway National Recreation Areas in 1972, Congress began to establish NPS units in major urban areas of the country. With each additional urban park—Cuyahoga Valley National Park in Ohio in 1974, Chattahoochee River National Recreation Area in Georgia, Lowell National Historical Park in Massachusetts, and Santa Monica Mountains National Recreation Area in California in 1978, the Timucuan Ecological and Historic Preserve in Florida in 1988, and Paterson Great Falls National Historic Park in New Jersey in 2009—the need for a variety of land protection tools grew and altered, necessitating new skills for the NPS managers and their community partners.

A major political reason for the NPS being pushed into the urban park business was the Nixon administration's elimination in 1974 of the well-funded Open Space Program in the Department of Housing and Urban Development. Initially a separate line item of Open Space grant funds in the Housing and Urban Development budget that supported parks in big cities, these park-specific grants, along with seven other "healthy cities" programs, were theoretically folded into a Community Development Block Grant program, but city park funding largely dried up after that.

In 1978, I worked closely with the House, Senate, and the Carter administration to support enactment of the NPS Urban Parks and Recreation Recovery Program (UPARR), a competitive big-city parks grant program. Congress felt that UPARR was an urban necessity because the state assistance block grant funds of the Land and Water Conservation Fund (LWCF) were not putting much funding into the largest cities' parks. Unfortunately, Congress chose to provide the revenue for UPARR from the General Fund appropriations rather than from the LWCF revenue that comes from outer continental shelf oil-and-gas leasing royalties paid to the federal Treasury. While UPARR was fully funded during the remaining years of the Carter administration, it was zeroed out in the Reagan administration by DOI Secretary Watt and only sporadically funded after that.

In 2002, while I was executive director of the National Recreation and Park Association (a society for park professionals, focused on local agencies), I also became a founding board member of the City Parks Alliance (CPA), which has become a powerful and effective advocacy and education nonprofit working for the multiple benefits—recreational, environmental, educational, and economic—of parks and open spaces in the big cities of America.

A major CPA success story has been its ability to convince Congress to establish a subaccount in LWCF, on the state assistance program side, for a competitive urban grant program, now called the Outdoor Recreation Legacy Partnership (ORLP). Informally established initially through the House Appropriations Committee's annual actions, with support from NPS leadership, ORLP was formally established in 2014, providing grants awarded by the NPS to cities of 50,000 or more in population. At present the need for NPS support of city parks is greater than the funding level of ORLP, in part because the annual amount available to ORLP is not established in law. Instead it is subject to annual decision by the appropriators, necessitating regular lobbying by CPA and its partner organizations. On the positive side, while ORLP funding has hovered around $12 million annually for some time, the FY 2021 Appropriations Act for DOI included $150 million for ORLP grants.

For each of its new urban NRAs, the NPS was given a minimum of core federal lands, authority to buy private lands or to receive donation of municipal lands, and statutory language directing it to work with local governments to expand land protection strategies, first through local actions in order to be successful. A key element common to all of these urban units is that, for the first time, Congress drew federal boundaries around acreages far larger than what was ever likely to be purchased by the federal government.

While the specific land protection tools vary among the NPS urban parks, one attribute common to all of them is the necessity for cooperation across governmental agencies at the federal, state, and local levels. Thus, in these urban units, Congress established the precedent of expecting NPS involvement in management of substantial nonfederal lands, recognizing that the NPS role would shift from simple land manager, as in more traditional units of the system, to intergovernmental collaborator, technical planning assistance provider, and ultimately to private land use regulator when locals fail to take on that role.

Each NPS urban unit attracts heavy public use for urban residents who cannot easily reach more remote national parks, and each has reduced the threat of over-dense development in already crowded communities. Each has provided successful urban "green space" conservation and/or preservation of historic structures; each has improved access to open spaces for urban dwellers, mostly through adaptive reuse of surplus or underutilized lands. The NPS commitment to city parks is as permanent as any federal program can be. Given sufficient time, each of these NPS urban parks will demonstrate the full potential of city parks as vital to the quality of urban life as well as to the economic engines that American cities have become.

The fifth partnership category includes those where the NPS must work with another federal agency in order to protect park resources and the larger landscape of which the park is only a part. The practice of working at the ecosystem scale, in cooperation with adjacent land managers like the BLM or the US Forest Service, started at Yellowstone with a draft "vision" document in 1990. Conceived by NPS Rocky Mountain Regional Director Lorraine Mintzmyer and US Forest Service Regional Forester John Mumma, the idea was that the area should be managed and coordinated as a larger ecosystem. The vision document recognized that the Yellowstone ecosystem supported migratory species, economically dependent communities, working forests and ranches, a vibrant tourism season, and an interdependent ecosystem. However, this vision was viewed as a threat to the extractive industry and as an overreach by the National Park Service. Both Mintzmyer and Mumma were forcibly reassigned from their jobs by John H. Sununu, then chief of staff to President George H. W. Bush.[2]

Perhaps the greatest competition between adjacent land managers has been between the US Forest Service and the National Park Service. The roots of this division go back to the days of John Muir, Gifford Pinchot, and President Teddy Roosevelt, when the idea of national parks and national forests was in its infancy. This story is better told by other authors but its legacy is still alive. This adjacency conflict is also alive and well in the Department of the Interior, where the BLM and the NPS are often at odds over extraction along park boundaries. This sort of competition and conflict is counter to, and inhibits success with, large landscape conservation, and must be addressed in the near future by an administration more willing to embrace the concept.

The sixth partnership category is the opportunity and responsibility to partner with Native Americans to comanage traditionally associated lands. For most of the twentieth century, the NPS has had a troubled relationship with

Native American tribes (including some not federally recognized) that are culturally affiliated with areas now national parks, or where a park's boundary
overlaps with a reservation boundary, or where the park simply overlays tribal
ancestral lands. In the late nineteenth and early twentieth centuries, the NPS
even forcibly evicted remnants of the Miwok Tribe from Yosemite Valley. And
during the same period, the NPS destroyed the winter lodges of the Timbisha
Shoshone Tribe in Death Valley National Park. In a few cases, Congress drew
park boundaries overlapping already extant reservations, such as at Redwood
National Park and the Yurok Reservation, Badlands National Park and the
Oglala Pine Ridge Reservation, and Canyon de Chelly National Monument,
located entirely on the Navajo Reservation.

The reservation boundary of the Yurok Tribe of California follows the
Klamath River from its mouth on the Pacific Ocean upriver for fifty miles to
its confluence with the Trinity River. At the mouth of the Klamath, the Yurok
Reservation overlaps with the later-established boundary of Redwood National
Park. Many traditional cultural activities of the Yurok occur on lands at the
Pacific shore of the park, some on the reservation and others on park land.

In 2005, the Yurok Tribe hired me to support their efforts to develop both
a tribal park on their reservation and a cooperative management agreement
between the tribe and Redwood National and State Parks, given the overlapping park and reservation boundaries and the cultural heritage landscapes
important to the tribe all across the parks. Over many months and meetings
from 2005 to 2012, a tribal park committee of elders, the Tribal Historic Preservation Officer Dr. Thomas Gates, and I met with both NPS and California
State Parks senior leadership to seek consensus. Jonathan was the regional
director overseeing Redwood National Park and was supportive of this collaboration. The tribe and the parks signed a "Shared Values Statement" intended to clarify the intent of all parties to collaborate and adopt some elements of comanagement. The tribe did not seek hunting rights in the park
but did want to be able to harvest and dry fish on the park's beaches, where
the fish would cure better in the salt air of the Pacific, and to harvest certain
native grasses from within the park for traditional hats and other apparel.
The tribe also wanted be able to take certain storm-downed redwood trees in
the parks for whole-log canoe carving, a customary practice. Up to that time,
the park would only allow the tribe to take a redwood log if its fall blocked
a road, and then only a piece as long as the width of the road—which was
not long enough for their traditional canoes. After nearly a decade of work,
through three changes in park superintendents, the untimely deaths of both
the tribal chair and tribal executive director, we were not yet able to finalize a
formal comanagement agreement. Nevertheless, the desire remains as work
for the future.

Mouth of Klamath River in both Redwood National Park and the Yurok Reservation. Credit: Destry Jarvis.

In addition to the battlefields of the so-called Indian Wars, like Big Hole and Little Big Horn, Sand Creek and Washita, a few other NPS-managed national historic sites have been designated by Congress as native cultural sites. These include Pipestone in Minnesota, Poverty Point in Louisiana, Knife River Indian Villages in North Dakota, Effigy Mounds in Iowa, Hopewell Ceremonial Earthworks in Ohio, and numerous archeological sites in the

Southwest like Chaco, Mesa Verde, and Hovenweep. However, until very recently, the NPS did not make serious efforts to work with the native people who were culturally affiliated with these parks, either as interpreters to tell their own stories to park visitors or as management partners for sites or resources that are culturally important to tribes. Fortunately, things have begun to change for the better.

The NPS has tentatively begun work between culturally affiliated tribes and associated parks, including consideration of broader cooperative management partnerships. As Director Kennedy's special assistant for policy and legislation in the 1990s, I worked with Congress to get a law passed to return a parcel of land within Death Valley National Park to the Timbisha Shoshone Tribe for residential use—land that they had occupied "illegally" since well before the park was established. A more recent accomplishment was the finalizing of a regulation in 2016 (by Jonathan) that allows for traditionally associated tribes to collect plants and plant materials for traditional uses from culturally affiliated parks. This effort took over fifteen years to complete but has built new relationships between parks like Great Smoky Mountains National Park and the nearby Eastern Band of Cherokee Indians. Much more should be done in this regard. The NPS should set a policy course to achieve optimum comanagement of parks with culturally affiliated tribes.

JONATHAN

My first foray into the politics and policy of boundary issues was in my second job, as a GS-5 park ranger at Prince William Forest Park in 1978. Once an eroded landscape, it was transformed by the Civilian Conservation Corps into a Recreation Demonstration Area. Today it is a lovely eastern deciduous forest of 16,000 acres, stretching over the piedmont-tidewater break with two small streams that feed into the Potomac River. It is bounded on two sides by the sprawling Quantico Marine Corps base.

As a ranger, I was interested in the condition of the natural resources of the park and, in particular, the significant volume of sediment that was pouring into the Quantico Creek from the upstream watershed within the Quantico Marine Corps base. With the assistance of Russ Davis, a biologist from the regional office, we installed a solar-powered sediment monitoring system at the park border with the base and began collecting water samples with unusual quantities of sand and silt. Off duty, I hiked through the Marine base and discovered that the Marines had permitted a commercial logging operation. The logging company was regularly dumping slash into and driving equipment through Quantico Creek, clearly a violation of standard forest practices. With some research on who might have jurisdiction, I contacted the Army Corps of Engineers and reported the situation. After a field visit, the Army Corps sent a cease-and-desist order to the Marine Corps for violations of the Clean Water Act. To say the least, when the senior leadership of the NPS learned that a GS-5 ranger had fueled a long-standing feud between the Army Corps of Engineers and the Marine Corps, I found myself sitting in the office of the deputy regional director of the NPS. That was my first meeting with the extraordinary Bob Stanton (who later became the fifteenth director of the NPS), who said that he appreciated my goals but not my methods. Bob saved my ass and we have been friends since that day. For the rest of my career, I have been paying attention (with improved methods) to what is happening outside the parks as well as inside.

Some thirty-five years later, I was sitting in a conference room in Washington, DC, discussing proposals for large solar energy projects for the California desert near Mojave National Preserve and Joshua Tree National Park. While I, as well as all the NPS, supported a shift to renewable energy, we also recognized that such projects have resource impacts to the water table, to wildlife, and to the landscape. We were seeking some protection for the nearby park units. As director, I had my team of resource specialists present, and the Bureau of Land Management director had his team along with the BLM's Nevada state director, Ron Wenker. Bert Frost, the NPS associate director for Natural Resources, stated our position, that the further away from the preserve the better. BLM State Director Wenker stated, very succinctly, that "the NPS needs to get used to the fact that the BLM is your trashy neighbor." That made us all laugh, but the implications were serious.

The Department of the Interior (DOI) is not a monolith, nor a cabinet of the executive branch with a uniform mission, such as the Departments of Education or Transportation. Instead, it is an aggregate of bureaus with inherently conflicting mandates that include the conservation of our national parks, wildlife refuges, endangered species, wilderness areas, and public lands that represent 20 percent of the land base of the United States. The DOI also includes the Bureau of Indian Affairs and carries out the federal government's trust responsibilities to over 500 federally recognized Native American tribes. The DOI is also responsible for permitting the extraction of oil, gas, and coal by private companies on federal lands. The DOI permits the transfer of rights to mine gold, silver, copper, uranium, and other hard rock minerals from federal lands, charging $2.50 to $5.00 per acre, a fee that has not changed much since 1872.

The DOI permits ranchers to graze their cows and sheep on public lands for $1.35 per AUM (animal unit month: which is one cow and one calf; 2020 rates). The DOI still authorizes "chaining": two bulldozers drive across the desert lands with a chain between them, mowing down native juniper trees to allow grass to grow for cows. The DOI authorizes below-cost timber sales, where the government subsidizes the building of roads for timber companies. The DOI permits the drilling of oil on the continental shelf and in the Gulf of Mexico, including the Macondo Well that blew up and resulted in the disastrous 2010 Deepwater Horizon Gulf oil spill. On the flip side, the DOI, through the National Park Service, is the keeper of the most important historical sites in the United States, including the Statue of Liberty, Flight 93, the Lincoln Memorial, and dozens of presidential homes and libraries. The DOI enforces (or not) the Endangered Species Act. Conflicts between these mandates and their parent bureaus is a form of internecine warfare, mostly conducted in the halls of the Department of the Interior (or, when it involves the US Forest Service, the Department of Agriculture). Getting the department, which is politically led, to think collectively about its responsibilities is a challenge that is rarely achieved.

In 1991, Dr. John Freemuth wrote a seminal work entitled *Islands under Siege: National Parks and the Politics of External Threats*. Dr. Freemuth, who passed away in 2020, characterized the external threats as enabled by "ineffective legislation, inept implementation, and the potent political power of pro-development forces."[3] Not much has changed, though there are glimmers of hope.

Sometimes, in the field far from Washington, there are opportunities to resolve border issues. When I was the superintendent of Craters of the Moon National Monument, I inherited an old problem: the northern boundary of the park had been drawn along section lines. Those lines crossed in and out of the park watershed, which contained the primary spring-fed potable water source for the campground, park residents, and visitors. Adjacent lands were leased for cattle grazing under the authority of the Bureau of the Land Management. As a consequence, cows were grazing and pooping directly above the park water source. There had been many efforts to adjust the boundary to reflect the watershed contour, but the effort had no traction in Congress. The park was faced with installing a very expensive filtration and water purification system on the spring to avoid cow-waste contamination. Fortunately, soon after I arrived, I got a phone call that started like this: "Howdy Jon, this is Leroy Cook from the Bureau of Low Morale." I knew instantly that I was going to like working with Leroy. So, I made a proposal. "Let's get a couple of horses, ride the northern border between the NPS and the BLM, and lay out a new fence line and just agree this would be the new park boundary." We made a day of it, had a great ride, and, later that summer, I hired a Student

Conservation Association crew to build a new fence along the line Leroy and I had identified. Legislation came years later that codified our change, but, in the meantime, cows were not in the watershed.

There are dozens of these stories that reaffirm how parks are adversely affected by outside activities. When I was working at Crater Lake National Park, we were confronted with the prospect of geothermal drilling on the park boundary to use the heat of the Mount Mazama volcano to make electricity. A nasty fight ensued with the developer, ultimately requiring Congress to designate Crater Lake as a geothermal resource that was protected from impact. A few years later, as the chief of resources at North Cascades National Park, I knew the ecosystem where I hoped to restore a grizzly bear population included US Forest Service lands as well as lands in Canada. These great beasts had once roamed these mountains but had been exterminated by hunters and government trappers. To accomplish their return, we would need support from the State of Washington, the US Forest Service, multiple counties, and the public. The concept was controversial. The US Forest Service, in particular, was worried about the impact on its ability to harvest trees. In one memorable meeting that I attended with the forest supervisor of the Mount Baker Snoqualmie National Forest, he complained he had enough problems with spotted owls, old growth, Indians, salmon, and added he didn't need any "goddamn grizzly bears."

At Mount Rainier National Park, where I served as superintendent, the park boundaries could be seen from space, because the adjacent timber lands of the US Forest Service and private forestry were cut right up to the section lines of the park boundary. And there were two spots on the boundary where the activities of the US Forest Service slopped over into the park. I worked out a land exchange with the local forest supervisor, but when we sent it to Washington, it died. And of course, there is fire, which knows no boundaries. In one of the most devastating fires in California's history (prior to the Camp Fire of 2018 and the Tubbs Fire of 2017), the Rim Fire started in August of 2013 from an illegal campfire on the Stanislaus National Forest and burned 257,314 acres, of which over 70,000 acres were in Yosemite National Park. In the national forest, heavy fuels were present from years of fire suppression, and the results were devastating. In Yosemite, which had spent years practicing prescribed burning, the Rim Fire had little environmental effect and was, for the most part, beneficial. Regardless, fire does not respect boundaries.

Over my career, I have worked with many wonderful professionals in the US Forest Service who are as committed to their mission as I was to mine. They revealed to me that there is a deep, well-ingrained resentment of the transfer of US Forest Service lands to the National Park Service because, for the rank and file, it is perceived as a loss and a failure of their stewardship.

I was told by a USFS friend that the chief of the Forest Service tasked one of his senior staff to "never let the National Park Service get another acre." When I came to Washington as NPS director in 2009, I saw a unique opportunity for better collaboration at the ecosystem scale between the four land management agencies. For the first time in history, all four had career directors. Sam Hamilton was the director of the Fish and Wildlife Service, Tom Tidwell the chief of the Forest Service, Bob Abbey director of the BLM. All four of us had worked our way up through the field and came to Washington to lead our respective agencies under the Obama administration. If ever there was a time to bury old hatchets, it was then.

A powerful incentive for working at the ecosystem or landscape scale is the realization that anthropogenic climate change is impacting our parks and public lands in new and significant ways. Scientists who study climate change noted that species would be forced to move to new refugia. In exceptional research, scientists at the University of California, Berkeley, revisited data sites in the Sierra mountains recorded by Joseph Grinnell between 1914 and 1920, and the new results revealed that species were indeed on the move to new habitats. If parks and other protected areas are considered islands, what lies between those islands? Wildlife seeking passage encounter barriers such as highways, fences, cities, ranches, farms, reservoirs, powerlines, energy developments, mines, and so on. So how could we begin to create connectivity across large landscapes? The first step was to identify the gaps and start filling them in, using funds from the Land and Water Conservation Fund.

The LWCF annual appropriation at that time was around $150 million, derived from the sale of fossil fuels from the outer continental shelf. The LWCF is allocated by Congress to the states and the four federal land agencies to use for land conservation, usually in the form of acquisition. Each of the four agencies gets a portion of these funds and seeks to buy lands that are essential to their agency objectives. For the NPS, which has over 2 million acres of private lands called inholdings inside national parks, the priorities for acquisition were those inholdings threatened with development. For the Fish and Wildlife service, critical wildlife habitat was prioritized, and for the Forest Service, it was specific forest types, such as longleaf pine. For the BLM, the priority was either adding missing parcels to their wilderness or landscape conservation areas, or access for recreation.

During the Obama administration, with support from the Secretaries of the Interior and Agriculture, the four directors gathered each year to pool a portion of our LWCF allocation and focus on large landscape conservation and connectivity. Tasking our technical staff with evaluating land acquisition priorities within a particular ecosystem, we would then create a joint list and the funds would go to whichever agency had the authority to buy the land or easement. Our collaboration worked to improve the integrity of the northern Rocky Mountains, the watersheds of the Chesapeake Bay and of the Everglades, the desert lands of the Southwest and California. This was a huge accomplishment and, frankly, the way I believe the public expects the agencies to work together. This is hard work and requires dedicated professionals and strong political leadership.

A second strategy for large landscape conservation was to focus collaborative effort around a charismatic species such as the pronghorn antelope. These beautiful animals migrate each year from winter range in the Green Valley of Wyoming over a hundred miles to summer range near Jackson Hole and Grand Teton National Park. Using the best science and working with ranchers, nonprofit organizations, the US Forest Service, and the NPS, barriers along the route were removed or modified so that this age-old migration could continue. Science has also shown that wildlife will use underpasses and culverts to cross major highways; engineered wildlife crossings have been successful in places like Banff National Park in Canada. The model is now being developed in the United States, where wildlife crossings are being designed for the Santa Monica Mountains National Recreation Area to accommodate mountain lions and other species.

Coordination and collaboration became more and more the way to operate. We were able to hammer out agreements with the BLM to (temporarily) protect parks like Chaco Canyon from adjacent oil and gas drilling that threatened this World Heritage Site. We reached agreements that would protect Death Valley's endangered pupfish from energy projects that would draw water from the aquifer. The directors of the BLM, the USFS, and I agreed to recommend to the Secretary of the Interior that 1 million acres around the Grand Canyon be withdrawn from leasing for uranium mining, an activity that would have threatened the springs of Grand Canyon National Park and potentially the Colorado River. The BLM agreed to move large solar arrays away from Joshua Tree National Park. In an excellent example of how the relationships between competing interests within the Department of the Interior are supposed to work, the Bureau of Ocean Energy (the new agency that leases lands and waters from 3 to 200 miles out) agreed to push the lease area for wind generators out so far that they would not be seen from Cape Hatteras National Seashore, even from the top of the lighthouse! This preserved a view of the ocean for hundreds of thousands of visitors. The Bureau of Reclamation agreed to open the floodgates of Glen Canyon dam to restore the eroded beaches and ecosystem of Grand Canyon National Park. These accomplishments came via hard work by professionals in all the agencies and strong political leadership at the top, which sometimes forced the opposing sides to the table. Most have been reversed during the Trump administration, whose "energy independence" mandate has become the top priority over everything else on the federal public lands and waters.

Unfortunately, especially in Washington, old habits and animosities are hard to kill. A good example was when President Obama was considering establishing a new national monument in the San Gabriel Mountains of California. This was a US Forest Service area that the National Park Service had extensively studied, at the request of Congress, as a potential new unit for the park system. Already managing the nearby Santa Monica Mountains National Recreation Area, the NPS saw the potential for a conservation and recreation area very near the highly diverse population of Los Angeles. While there was great local support in the surrounding communities for a new protected area designation, the political question of which agency would manage it was not yet settled.

One of the components of the competition between the USFS and the NPS is budget. The NPS receives much greater funding, per acre, than does the US Forest Service. In fact, the allocation from Congress goes directly to each park for its operation, whereas the Forest Service budget is much more dependent on factors such as timber sales. Since the new monuments do not generate revenues from extractive activities, they have been notoriously underfunded in the Forest Service, which results in unmaintained restrooms, trails, signs, and other amenities and few rangers if any. As the momentum was building for a presidential proclamation to make some 340,000 acres of the Los Angeles National Forest into the San Gabriel National Monument, I set up a meeting with my friend and colleague Tom Tidwell, the chief of the Forest Service. Meeting in his office, where Tom still uses the desk of the first chief, Gifford Pinchot, I made the case that this was the opportunity for a collaboration between the NPS and the USFS; together we could make this new monument a great success. If the NPS had a piece of the monument but not the whole thing, large enough to justify a budget request for us both, we could perhaps end a hundred years of infighting. I asked Tom to support transferring 20,000 acres of the proposed 340,000-acre monument to the NPS as a part of the proclamation. Tom, in the most friendly and gentlemanly way, said no. He said that the Forest Service had to prove that they could do this without the NPS. I think it was a major missed opportunity. The San Gabriel Mountains National Monument is operated by the US Forest Service and continues to struggle to meet the needs of conservation and access for the public, both due to lack of funding.

As Destry detailed above, the lands within the national park system are part of the traditional homelands for Native Americans. When Columbus landed on what would be later determined to be a new continent, it was already occupied. There were millions of people living here in thriving societies with governance, language, complex cultures, agriculture, architecture, and religion. Over the next five hundred years of war, disease, discrimination, assimilation, broken treaties, and genocide, native people would be removed from their homelands, including lands that would become national parks. In that today's national parks are the closest thing we have to the lands that existed prior to colonization, it is essential that the NPS develop meaningful relationships with native people who have traditional relationships with the lands now designated as national parks. My first meaningful experience working directly with native people was when I was negotiating the relicensing for the three hydroelectric dams within Ross Lake National Recreation Area in the mid-1980s. For three years, I met weekly with the representatives of five tribes who called the Skagit River home; they were determined to seek mitigation for the loss and impact to the fisheries and wildlife. I came to respect their deep sense of time as well as their patience and persistence against uneven odds.

At Craters of the Moon National Monument, in my first superintendency, it bothered me that we referred to the park's two types of lava as "aa" and "pahoehoe," names used by Native Hawaiians. I knew that the Shoshone people would have witnessed the lava flows since they had lived in the area for the last 10,000 years. Surely, they had names for the lava fields that would be more culturally appropriate. I reached out to the Shoshone Tribal Council and asked for representatives to come to the park and spend some time with me and the staff. Three elder women came to the park for the day, but when I asked if they knew Shoshone terms for the lava, they said that as children, they had been removed from their families and sent to schools where, when they spoke their language, the teachers would wash their mouths out with soap. So, they thought there were names for the lava, but they no longer knew them.

In Alaska, I worked daily with the Ahtna, Eyak, Tlingit, Doyon, and other Native Alaskans in helping to preserve their rights for subsistence hunting and fishing. To me, their thousands of years of association with the lands in the park made the park stewardship richer and more sustainable. At Mount Rainier, I made a point to visit the tribal councils of the six affiliated tribes—the Nisqually, Muckleshoot, Cowlitz, Yakama, Squaxin Island, and Puyallup—and invite them back to the park. At my meeting with the Yakama Nation, an elder said to me that the last time a contingent of tribal members had come to the park, a ranger had run them out. I knew this story, as it happened around 1915, so I apologized and said, "those rangers don't work at the park anymore and this time you will be welcome." It was through these discussions that I became aware that the NPS regulations on the collecting of plant materials by native people was written in such a manner that could easily be viewed as biased if not racist. While the general public is allowed to collect mushrooms, fruits, nuts, berries, and unoccupied sea shells, native people are prohibited from collecting anything unless it was pursuant to a treaty. This seemed an opportunity for a significant improvement in relationships, if managed properly, in collaboration with the leadership of tribes who are traditionally associated with a specific park. I made a commitment that I would change that regulation. In July of 2016, fifteen years after I began this effort, as director of the National Park Service, the final rule was completed:

> The National Park Service is establishing a management framework to allow the gathering and removal of plants or plant parts by enrolled members of federally recognized Indian tribes for traditional purposes. The rule authorizes agreements between the National Park Service and federally recognized tribes that will facilitate the continuation of tribal cultural practices on lands within areas of the National Park System where those practices traditionally occurred, without causing a significant adverse impact to park resources or values. This rule respects those tribal cultural practices, furthers the government-to-government relationship between the United States and the tribes, and provides system-wide consistency for this aspect of National Park Service–tribal relations.[4]

There is much yet to be done in this area. Unlike other nations, such as Canada, the United States has yet to establish a national park owned and operated by a Native American tribe. As director, I pursued an opportunity with the south unit of the Badlands National Park, converting it to a tribal national park, operated by the Oglala Sioux Tribe. It was once a part of their reservation but was removed to be used by the military during World War II as a practice bombing range. After the ordnance was removed, it was then transferred to the National Park Service as part of Badlands National Park. It should be returned to the tribe and operated by them as a national park, with the NPS providing technical assistance. Politics stalled this effort. This act would go a long way toward establishing trust between Native Americans and the NPS.

Our Decades of Park Experiences Make Clear That the Parks Won't Survive, Much Less Flourish, without Partners

The debate about impacts to national parks from activities on adjacent lands has raged for decades and needs to have clarity provided by legislation in Congress. Within the Department of the Interior, the issue came to a head in 1998, when Secretary of the Interior Bruce Babbitt asked for a formal opinion of the solicitor regarding the Secretary's discretion to permit extractive activities on federal lands outside of national parks, knowing that these activities would potentially impact the park. The solicitor, John Leshy, wrote a long and well-researched memorandum, often referred to as the Doe Run opinion, that concluded:

> In sum, I believe the text of the 1978 (Redwood) Amendment and the other legal considerations discussed in this section support the conclusion that the Organic Act as amended in 1978 does have application to the Secretary's exercise of his authority over activities taking place outside the boundaries of park units.[5]

The implications of this legal opinion are that the Secretary does not have the authority to "derogate" (allow to be degraded) national park resources through decision-making to develop resources on BLM public lands that are adjacent to national parks. Few if any Secretaries have heeded this advice. Instead they attempt to "balance" responsibilities by picking winners and losers or to openly side for development while actively suppressing the objections of NPS professionals.

As mentioned in chapter 3, the National Park System Advisory Board recommended that the goal of resource management of the National Parks should be:

> . . . to steward NPS resources for continuous change that is not yet fully understood, in order to preserve ecological integrity and cultural and historical authenticity, provide visitors with transformative experiences, and form the core of a national conservation land- and seascape.[6]

We know that park resources are highly vulnerable to the activities that take place within a larger landscape, beyond park borders. We also know that climate change is pushing species out of parks into other lands, where their conservation is threatened. With limited authority beyond artificial boundaries, for the present the NPS must rely on partnerships, cooperation, agreements, and a larger vision to be ultimately successful. Neither lack of desire to work at that scale nor capability to participate as a value-added partner are the primary impediments. Political interference is.

Politics and the seesaw administration of the Department of the Interior are at the heart of the conflict that has endured for decades. The administration of President Donald Trump, in particular, has shown the vulnerability of professional management of our national parks, as career public servants with decades of experience working with gateway communities, nonprofit organizations, businesses, other federal agencies, and conservation organizations were unceremoniously tossed from their jobs or forced into retirement. Park superintendents were directed, with serious consequences if they disobeyed, to remain silent on activities outside park boundaries, regardless of their threat to the integrity of the national park.

While Congress has at times attempted to legislate better protection for the national parks, supportive administrations have put in place science-based policies, and the courts have often ruled in favor of park protection, all of these are ignored, unwound, or reversed during a hostile administration. The only long-term solution that we see as viable is to decouple the National Park Service from the Department of the Interior and make it independent.

SIX

INTERFERENCE IN THE MISSION

The Organic Act of 1916, establishing the National Park Service, is quite clear in its intent:

> The service thus established shall promote and regulate the use of the Federal areas known as national parks, monuments, and reservations hereinafter specified by such means and measures as conform to the fundamental purpose of the said parks, monuments, and reservations, which purpose is to conserve the scenery and the natural and historic objects and the wild life therein and to provide for the enjoyment of the same in such manner and by such means as will leave them unimpaired for the enjoyment of future generations.[1]

Congress reaffirmed this in the National Park Service General Authorities Act in 1970:

> . . . these areas, though distinct in character, are united through their interrelated purposes and resources into one national park system as cumulative expressions of a single national heritage; that, individually and collectively, these areas derive increased national dignity and recognition of their superb environmental quality through their inclusion jointly with each other in one national park system preserved and managed for the benefit and inspiration of all the people of the United States.[2]

And again in 1978, in an amendment to the General Authorities Act, attached to the act expanding Redwood National Park:

> The authorization of activities shall be construed and the protection, management, and administration of these areas shall be conducted in light of the high public value and integrity of the National Park System and shall not be exercised in derogation of the values and purposes for which these various areas have been established, except as may have been or shall be directly and specifically provided by Congress.[3]

When one combines these three statutes, it is clear that it is not just expected but required that the National Park Service manage by "such manner and by such means as will leave [the parks] unimpaired for the enjoyment of future generations." This is what every employee of the National Park Service refers to as "the mission." The National Park Service and its advocates view any and all actions by politicians, presidents, developers, or adjacent land managers that may harm this mission as interference.

During the four years of the presidency of Donald J. Trump, 2017–2021, we witnessed the most systematic attack on the environment of any administration in our experience. While some past administrations ignored basic environmental laws and policies, this one was openly hostile and used its time in office to reverse a generation of actions that protect our air, water, and environment. The National Park Service did not escape this attack. In 2020, we participated in a team effort to analyze, curate, and prioritize the environmental harms of the Trump administration and provide them to the incoming administration so that they could be reversed. Specific to the NPS, the Trump administration's harms included:

- Leaving the NPS director position vacant for four years
- Forcing the reassignment and/or retirement of almost every senior career leader in the NPS
- Rescinding Director's Order 100, the new policy on climate change adaptation
- Opening up oil and gas drilling adjacent to Chaco Culture National Historical Park
- Authorizing hunters to kill female bears and their cubs, and wolves and their pups, in their winter dens inside Alaskan parks
- Threatening NPS scientists with discipline if they spoke publicly about climate change
- Removing references to climate change in park management plans

- Building the Mexican border wall through the entirety of Organ Pipe Cactus National Monument, which destroyed wildlife habitat and cross-border wildlife movements, drained fragile desert springs, caused huge erosion problems, and damaged important indigenous cultural landscapes. In perpetrating this clearly illegal resource impairment, no NEPA or NHPA section 106 analysis was done, the omission justified in the name of an illegal alien "emergency."
- Reducing the size of Grand Canyon-Escalante and Bears Ears National Monuments by 80 percent (which also impacts adjacent NPS parks)
- Allowing uranium processing adjacent to the Grand Canyon National Park
- Abolishing all NPS advisory committees and refusing to convene the NPS Advisory Board
- Proposing, each of their four years, to cut the NPS budget and personnel ceiling
- Making political use of several NPS units for live campaign events, including Mount Rushmore, Gettysburg, and Fort McHenry

Of course, the Trump administration was just one more in a long line of hostile teams in the department. Ronald Reagan's Secretary James Watt stated that if there was a problem with the NPS he would "change the policy or the person," and he issued a moratorium on all land acquisition for the parks. George W. Bush's Secretary Gale Norton prohibited park rangers from working with schools, stating that NPS environmental education was "mission creep."

Over the course of our combined experience, we have encountered many examples of interference in the mission of the NPS. For example, in 1983, DOI Secretary Watt, over strenuous internal NPS objections, permitted the Jackson Hole, Wyoming, commercial airport to expand its runway into Grand Teton National Park, bringing many more jets over the park's airspace, adversely affecting both wildlife and the quality of the visitor experience. The most egregious actions are those that use the latest government initiatives like "energy independence," or budget and personnel cuts, to control professional decisions and change priorities. For the past fifty years, these Republican administrations have clearly favored policies of mining, extraction, commercial timber cutting, and so on, which are the antithesis of national park preservation. In such instances we have had to invoke the NPS Organic Act's explicit authority for NPS professional staff to manage parks "in such manner and by such means as will leave them unimpaired for the enjoyment of future generations" to thwart the harm, though, far too often, unsuccessfully.

DESTRY

Every Republican administration since Richard Nixon's second term has sought, some more effectively than others, to overtly politicize the NPS and radically shift the agency's mission. They have sought both to foster the wrongheaded idea that the first priority for NPS is visitor use and associated commercial development, and to prevent professional NPS managers from objecting to adverse impacts to park natural and cultural resources, whether caused by inappropriate visitor uses or by adjacent developments outside of park boundaries, many caused by other federal agencies or federally permitted projects.

These radical Republican policy swings, insofar as internal NPS policy is concerned, have sought changes to suggest that NPS must balance preservation and use, whereas the correct interpretation of the Organic Act's mandate is that resource preservation is the priority and only that use which is compatible with preservation of the parks unimpaired is permissible.

Some have gone further, as with the Bill Horn versus Bill Mott saga (see chap. 3), to assert that visitor use is the dominant purpose of the parks. As tools to implement these changes, beyond the *Management Policies* manual, they have imposed new visitor fees and devoted the revenue to expanded access and infrastructure; fostered new roads, parking, and commercial services; and contracted out more park functions that have been and ought to remain inherently NPS tasks.

Each of these Republican administrations has inserted senior political appointees to control the NPS, either as a political deputy director, as senior staff to the assistant secretary for Fish, Wildlife and Parks, or directly as acting director. None of these required Senate confirmation. Reagan's DOI had Rick Davidge, who had been staff in the National Park Inholders Association, and Mary Lou Grier, who had previously revived the Texas Republican Party. George W. Bush had Paul Hoffman, formerly director of the Cody, Wyoming, Chamber of Commerce. In addition, throughout the Reagan, Bush, and Trump years, P. Danny Smith, a former lobbyist for the National Rifle Association, served as a political controller of the NPS in multiple positions: as deputy assistant secretary, NPS assistant director for Congressional Affairs, deputy director, and acting director. Unlike most political appointees who come and go, Danny stayed long enough to fully grasp how to keep his thumb on the policies of the NPS, from the profound to the petty, and, more often than not, contrary to the NPS mission. As a general matter, for all of these political appointees, their policy demands were for increased commercial developments, increased visitor fees, minimal or no science to inform actions, and strong support for recreational use of the parks "trumping" resource preservation.

In 1978, I worked closely with the House Interior Committee staff to successfully enact an amendment to the NPS General Authorities Act that requires the NPS to "determine and implement carrying capacity" for every unit of the system. Without a doubt, public use of the parks is a good thing, but that must be professionally managed by the NPS, especially when the numbers and types of use could otherwise impair park resources and the quality of other visitors' experiences. There is a well-documented correlation between park use and public support for the NPS, including within Congress. That said, there are upper limits to visitation in parks and, to date, the NPS has failed to carry out this mandate, at least in part due to the pressure from politicians representing gateway communities and tourism businesses dependent upon robust park visitation.

One major issue that has tormented the NPS throughout these past forty-eight years is management of concessions—the hospitality industry contractors who operate park hotels, restaurants, gift shops, and associated facilities and visitor services. In the wide swings of the pendulum between Republican and Democratic administrations, no issue has more dramatically revealed the radically opposite interpretations of the NPS Organic Act than concessions management—visitor use/recreation versus resource preservation.

Congress enacted the seriously deficient Concessions Policy Act of 1965 to authorize concession contracts and the development of the rules to manage them. The act's four big flaws were that 1) it offered incumbent contractors virtually perpetual renewal of their thirty-year contracts as long as they performed satisfactorily, but it did not require annual performance reviews by the NPS; 2) it guaranteed them a right to make a profit off park visitors; 3) it established a compensable possessory interest in concession facilities built by these contractors, the value of which appreciated like regular real estate over the length of their contracts; and 4) there was no requirement for an existing concession company to get prior approval from the NPS to sell its business to another operator. One provision of the 1965 act that had great merit, but was largely ignored by the NPS until the late 1970s, required that the concessioners' right to make a profit was tempered by a requirement that it be "consistent with satisfactory service to the public."

In 1975 I was for the first time involved in a lawsuit against the NPS. On behalf of the NPCA, we sought public disclosure of concession financial data (profit and loss) under the Freedom of Information Act. Our main objective was to publicly disclose these highly profitable businesses along with the comparatively low "franchise fee" they paid for the privilege of operating, noncompetitively, in our national parks. Generally concessioners paid 1/2 of 1 percent of net profit, which was deposited in the General Fund of the US Treasury and not used as part of the NPS budget. The NPS concessions management office was even run by a former concession company manager, which was just too much of a conflicted arrangement for the NPCA.

During the Carter administration, the first serious efforts to reform concession management were undertaken. Following years of complaints about the poor service by the concessioner in Yellowstone National Park, the NPS hired a concession manager from outside the service, L. R. "Buddy" Surles, to review the performance there. Buddy had been state park director in Arkansas under Governor Dale Bumpers, hired to reform and revitalize a poor concessions system there. After he successfully completed that Yellowstone performance review, the incumbent concession contract was terminated, and the contract competed for the first time in decades. Based on that success in Yellowstone, the NPS brought Buddy to headquarters as the new chief of concessions. He quickly instituted urgently needed reforms with the full support not only of the NPCA and myself but of NPS Director Bill Whalen as well as his old boss, Dale Bumpers, by then US senator from Arkansas. For the first time, the NPS hired new staff, both in parks with large concessions and in NPS regional offices, who were trained in hospitality management and accounting and who had not come from prior employment by NPS concession companies.

Unfortunately, in 1981, among his very first actions as Reagan administration DOI Secretary, James Watt summarily transferred Buddy Surles out of NPS headquarters into a much lower-level, nonconcession job in Denver—at the specific request of the concessioner's trade association, the Conference of National Park Concessioners (later renamed the National Park Hospitality Association). Watt then installed another former industry concession manager, the son-in-law of the owner of one of the multipark concession companies, as NPS chief of concessions, and any chance of reform was squashed. Of course, throughout this political hatchet job, on behalf of the NPCA and others, I strongly protested to the Secretary, on the Hill, and in the news media, but to no avail.

During the George H. W. Bush Administration, DOI Secretary Manuel Luján and NPS Director James Ridenour became alarmed that Music Corporation of America (MCA), the owner of the Yosemite concessioner Yosemite Park and Curry Company (YPCC), had been purchased by a Japanese corporation, Matsushita. Luján argued very publicly that no national park concession should be owned by a foreign company. Consequently, he forced Matsushita to agree to divest itself of the concession company. The NPS did not have funds to buy out the concessioner's possessory interest, which the company estimated at $200 million. Instead, Director Ridenour convinced National Park Foundation Executive Director Alan Rubin to buy out the concession contract. Since the NPF had no funds for this purchase either, Luján and Ridenour convinced MCA to loan the NPF $49.5 million at 8.5 percent interest and to sell at this bargain price. This deal in turn allowed the NPS to solicit bids for the concession contract of a major national park for the first time with no possessory interest to be paid by the winning bidder. On this key point and to induce more competition, it is notable that the contract bid was won by Delaware North Companies Parks and Resorts (DNC), which then bought the concession from the NPF at a much reduced cost in the absence of any possessory interest.

Between 1982 and 1993, Senator Bumpers introduced numerous concession reform bills—all with NPCA endorsements—but, strongly opposed by the Reagan and George H. W. Bush administrations, they did not pass. Fortunately, during the Clinton administration, while I was assistant director of the NPS, an opportunity for bipartisan concession reform emerged. Senator Craig Thomas, a former Wyoming hotel owner, became chair of the Parks Subcommittee, with Senator Dale Bumpers as the ranking minority member. Together, with their staff members Tom Williams, David Brooks, and Dan Naatz, working with other NPS staff and myself, we reached agreement on a major reform bill that was enacted as the NPS Concessions Management Act of 1998.

Among the major reforms of the 1998 act, the new law limits NPS concession contracts to a ten-year standard term, with an option for the NPS to go to twenty years when major new construction investment by the company is authorized by the agency. The preferential right of contract renewal for incumbent companies was repealed, along with a prior requirement that any new service needed by the NPS would be contracted to the existing company. Thus, every contract about to expire had to be competed on a much more level playing field. The prior ownership value, possessory interest, was required to be converted to a "leasehold surrender value" that was no longer tied to standard appreciation of real estate.

Another major provision of the 1998 Act is the requirement that the NPS determine, prior to soliciting bids for a concessioner, that the proposed commercial visitor service function is "necessary and appropriate" for proper management of visitor use in that park. For example, a park located close to a population center, or one with a gateway community at its boundary, may very well not need to have parkland taken up by a hotel or restaurant, services that can be provided to park visitors from outside its boundary and thus not warranting an NPS contract. For example, hugely popular national parks like Acadia and Great Smoky Mountains do not have hotels inside the park, because they each have great visitor services in surrounding communities.

The new law has led to a much stronger competition for new concession contracts. Notably, DNC was the only concession company to support enactment of the 1998 Concession Management Act, which expressly endorsed the need for instilling competition in the award of concession contracts. Today, there are typically five or six concession companies that fully consider bidding on every contract up for renewal or replacement.

When the George W. Bush Administration came into office in 2001, it launched another major assault on the professional integrity of the NPS workforce. A major new initiative required all agencies to "competitively source" federal jobs wherever possible. A team of political appointees at the Office of Management and Budget (OMB) as well as within the DOI—which excluded NPS and other career staff—reviewed NPS jobs, tasks, and functions. It determined that nearly 2,000 GS positions could be eliminated and offered through contract to the private sector, although each NPS employee would be allowed to compete for his/her own job.

The Rockefeller Family Fund hired me in 2003 to oppose these and other adverse privatization impacts on the NPS. It quickly became clear that DOI Secretary Gale Norton's push to impose these requirements was driven by a dislike/distrust of the NPS civil servants but cloaked in a stated intent to reduce federal costs.

What competitive sourcing and privatization advocates like Secretary Norton failed to understand, but for which there is ample social science data, is that the most important factors in increasing job productivity are personal satisfaction and dedication to the mission of the NPS. The mere process of requiring NPS employees to compete for their own jobs would have destroyed that job satisfaction and demoralized employees to the point that any savings created by the competitive process would be lost.

Of greatest importance for an agency like the NPS was the failure of the political appointees in OMB or DOI to comprehend the sort of multitask, multiskill knowledge and ability that most NPS employees in the field possess and regularly utilize. It is not so simple, as if the NPS could replace a GS maintenance employee with a contracted plumber or carpenter or electrician. An NPS maintenance employee could be a carpenter in summer, a firefighter in fall, and a snowplow operator in winter. No private company could hire equivalent workers, much less compete against this type of multitasking requirement on a basis that would reduce the NPS staffing budget.

It took the NPS, along with its congressional allies and numerous NGO supporters, which I helped to organize and rally, nearly three years to force the Bush administration to drop virtually all the NPS from the requirement of competitive sourcing. Little was accomplished by this initiative, which was wrongheaded from the start, other than its negative impact on NPS morale.

Had the goals of the Bush administration been carried out, it could have been the most devastating and demoralizing event for employees in the entire history of the service. It would have changed the agency from a public service, mission-oriented, creative organization into a group of contract administrators, who would have had to negotiate with the private sector for the protection of the most sacred and revered symbols of our American heritage. No collection of contractors could ever have the depth of commitment to the NPS mission that the career employees of the NPS demonstrate every day.

Over these past forty-eight years, some of my biggest political fights with political appointees have been over the many adverse natural resource and visitor use impacts stemming from motorized vehicles—off-road vehicles running over sand beaches in national seashores such as Assateague Island, Fire Island, Cape Cod, and Cape Hatteras; snowmobiles in Yellowstone; jet skis on flatwater park lakes; fixed-wing and helicopter air tours over the Grand Canyon and Hawai'i Volcanoes; and motorized versus oar-powered rafts on park rivers.

All of these types of vehicular use in the parks cause noise and disturbance of wildlife habitats and use patterns; ORVs on seashore beaches disturb, even kill at times, nesting migratory birds, including the endangered piping plover, and nesting sea turtles. Until very recently, most motors on river rafts, jet skis, and snowmobiles were two-stroke engines, highly polluting and very noisy. In Yellowstone, snowmobilers referred to "bison ping-pong" as they drove in and out of the bison herds that made convenient use of the packed and hardened snowmobile routes. Some park concession operators, like the raft companies in Grand Canyon National Park, voluntarily converted all raft motors to four-stroke as soon as the technology was available, while the NPS had to force it in other places.

Low-altitude commercial air tours over the Grand Canyon, for another example, are noisy and disturb visitors as well as the endangered condors; their vibrations have damaged delicate ancient Native American structures inside the walls of the canyon. For more than twenty-five years, political appointees in both the DOI and the Federal Aviation Administration (FAA) have refused to allow the NPS to prepare air tour management plans, stemming largely from the FAA's stated attitude that they "do not want the air space map of America pockmarked with restricted air space over national parks."[4] As assistant NPS director in the 1990s, I argued with FAA senior managers that the NPS should have the responsibility to determine whether any such air tour overflights were necessary and appropriate, and if so where they could fly with minimal resource and visitor impacts. The FAA's only job ought to be to regulate those tour flights to be safe.

Democratic administrations during my career—Carter, Clinton, and Obama—have all come into office with their own goals and initiatives, some of which have altered NPS priorities. But none of these administrations have tried to dramatically cut the NPS budget or slash personnel ceilings, ignore science, politically rewrite *Management Policies*, deauthorize existing NPS units, or muzzle the professional voices of NPS leadership. In sharp contrast, Republican administrations—Nixon, Ford, Reagan, both Bushes, and Trump—to varying degrees, but consistently, have sought to politicize the agency, to ignore science and professional analysis, cut its budget, reduce its personnel positions, and shift its focus dramatically to recreation and visitor use.

JONATHAN

I was the newly minted regional director for the Pacific West when I got the call. I was to travel to Santa Barbara, California, and meet with the Bush administration's Department of the Interior Assistant Secretary Lynn Scarlett.[5] I had read what I could find on Scarlett and knew she was a conservative but also an avid outdoors person, which gave me hope. She was accompanied by her assistant Dan Jorjani, and we met in the back booth of a dark Santa Barbara restaurant. I felt like this might be a scene from *The Godfather*, where I am part of the meal. The impetus for this secret meeting was the NPS's special resource study for the Gaviota Coast, the stretch of land running north from Santa Barbara to Lompoc, one of the last undeveloped coastlines in southern California. Prior to our meeting, I had spent the day exploring the area and it really was spectacular and worthy of designation.

The resource study was clearly leading to recommendation for congressional designation as a new unit of the NPS, perhaps a national seashore, like Point Reyes or Cape Hatteras. While there was considerable support among the local conservation community and the surfers who wanted better access, there was also strong opposition from the local property owners. Lynn was there to support the property owners and was direct over dinner: she told me to kill the study. I suggested we let it play out to the end and just recommend against the designation, and she said no. As the regional director, I was in my first Senior Executive Service position, the highest level of career public service in the federal government. It is also the most politically vulnerable: the new Bush team could reassign me, "for the good of the government" and with just sixty days' notice, to anywhere in the United States (or Guam for that matter). I chose to kill the study. The following day, Assistant Secretary Scarlett announced at a public meeting of ranchers that the study had been terminated. I was left to explain this decision to a very unhappy planning team on my staff.

As regional director of the Pacific West, I was often called to Washington during the Bush administration to meet with the political appointees of the Department of the Interior on a variety of issues. This time it was to discuss a proposal regarding Lake Mead National Recreation Area. I was ushered into the office of Dan Jorjani, whom I had met back in Santa Barbara. Joining Jorjani was Derrick Crandall, longtime Washington, DC, beltway lobbyist for both the motorized recreation industry and the national park concessioners. Crandall wanted me to authorize a Nevada developer to purchase, in bulk, passes to Lake Mead so that he could give them away to potential investors in his development near the park. I explained that the passes are considered accountable property and there was a long process of accounting for each pass, reporting requirements, and regular auditing. At that point, Crandall became irate, jumped up, leaned into my face, and literally screamed at me, spittle flying and demanding that I approve this idea just as he proposed it. Jorjani actually intervened and asked Crandall to settle down. This was a first for me—being screamed at by a lobbyist when I refused to do his bidding—but it has become a common occurrence for many career employees who stand up to those who would do harm to the mission of the NPS.

In 2006, I was visiting briefly in an office in the National Park Service Headquarters in Washington, DC. My region included Death Valley National Park, where the park superintendent, J. T. Reynolds, had just been featured in an article in *Vanity Fair*[6] that openly challenged the Bush administration with its treatment of the national parks. Paul Hoffman, who had been prominently featured in the article, popped his head into the office, shook the issue of *Vanity Fair* in my face, and shouted "What are you going to do about J. T. Reynolds?" I said, "J. T. is a Texas A&M football hero, Vietnam veteran, African American park ranger, who looks damn good in his uniform. You want to take him on, go for it." Hoffman turned and stomped out.

The Channel Islands are the American equivalent of the Galapagos Islands. Channel Islands was first proclaimed as a national monument in 1938, then redesignated as a national park in 1980. Separated from the mainland millions of years ago, these islands off the coast of California developed their own unique environment, and the species of wildlife that somehow got there changed and adapted to this isolated existence. For instance, the Channel Island fox, a distant relative of the red fox, became smaller but was still the "top dog" on the island. In contrast, Santa Rosa Island, within the park boundary but still private property, was owned by a series of ranching families who introduced cattle, pigs, sheep, turkeys, and elk to the island with impacts devastating to the native flora and fauna. When bald eagles succumbed to the effects of the pesticide DDT, they were replaced by golden eagles that killed and ate the feral piglets but also the island foxes.

In 1986, the National Park Service purchased Santa Rosa Island from the private owners and began the process of restoration by removal of the nonnative species. When I became the NPS regional director in 2002, there were still pigs, turkeys, and elk on the island, and the private landowners were using whatever political power they had to remain there, well beyond any legal claim. Although we were able to remove the pigs by lethal means, the elk population remained. Backed by the former ranchers, the local congressman, Duncan Hunter, introduced legislation to turn the island into a private hunting ground for veterans. Vehemently opposed by conservationists, and with the support of Senator Dianne Feinstein, we were able to block the legislation that would have turned an amazing national park into a hunting ground. Even the veterans who were contacted found this idea to be objectionable and felt used by the politics. When Secretary Zinke visited Channel Islands National Park in 2017, he refused to meet with the park staff or scientists and instead invited the former island cattle ranchers to lead his tour. Upon return to Washington he instructed the acting director of the NPS to fire the park superintendent and make the boat captain who had ferried him to the island the new park superintendent. The acting NPS director, who was a career employee, refused to follow that order.

With each new Republican administration over the past thirty years, political appointees dust off a directive from the Office of Management and Budget called A-76, or "competitive sourcing." The policy theory behind this directive is that 1) the private sector can do just about anything better than government; 2) that the government should only do those things that are inherently "governmental"; and 3) that if there is something that the government is currently doing, it can be improved by competing it with the private sector.

This philosophy has a number of inherent flaws, and its implementation has had significant impacts on the National Park Service. First of all, the things that are best done by the private sector for the NPS—such as lodging, reservations, food, beverage, guides, outfitters, and equipment rentals—are already done by private contractors under concession contracts. The George W. Bush administration's approach to this fact was "Fine, so what else do you have that we can give to the private sector?" They directed each national park to assess its current operations and submit something for "competition" with the private sector. I was the superintendent of Mount Rainier National Park at the time and really could not see any part of the operation that could be handed over to a private company without seriously impacting the park operations and its stewardship responsibilities, or actually costing the American taxpayer more.

Mount Rainier, capped with glaciers at 14,400 feet, is a destination for mountain climbers from all over the world. For those who desire to attempt the taller peaks of the Himalayas, Rainier is the training ground. The park hosts some 10,000 summit attempts per year, with about 50 percent succeeding and an average of three deaths and multiple rescues. Typical climbers overnight at Camp Muir or Camp Sherman for an early start to the summit and are led by private guide services, such as Rainier Mountaineering Inc., that operate as concessioners.

The political appointees in the department suggested that we turn over the maintenance of toilets for the mountain climbers to the private sector. This is where it becomes kind of both fun and excruciatingly stupid: sitting in a conference room with a group of shark-suited political staff and explaining in detail that to pump, clean, and service a toilet at 10,000 feet elevation requires an individual with unique mountaineering skills, helicopter load certification, and willingness to sleep multiple nights in a tent on snow. As a part of the mountain team, they must also be available to assist on the frequent mountain rescues, which means they are fit enough to summit the mountain, self-arrest in a free fall, have emergency medical training, and survive in a storm. Oh, and they also need to know how to clean a toilet and, in this case, operate the solar system that melts the . . . stuff. When told by political appointees that there are plenty of companies that clean toilets, I said I had never seen a toilet service that could supply highly experienced mountaineers.

A significant number of park superintendents put up fights against this destructive policy, recognizing that carving off a piece of each park's integrated operations for the private sector would undermine the mission of park protection and public safety. In this case, our means was to stall until these people went away. As one snarky employee stated to the incoming political appointees, "We are the B team. We 'be' here when you got here and we will 'be' here when you are gone."

By pushing back, we stalled the process through most of the Bush years, but ultimately NPS Director Mary Bomar relented under extreme pressure and offered up the entire human resources function of the NPS. Its highly professional specialists, who had been in a distributed system scattered across park organizations and totally integrated into park operations and needs, now had to defend their jobs against the private sector companies that believed it could all be done in a central office somewhere in the Midwest. While the NPS won the competition against private sector bidders, the agency was forced to implement the proposed structure, even though we knew it would not work. It was an unmitigated disaster. Many fine professionals left in frustration, and the National Park Service human resources function has yet to recover. Positions that used to take a few months to announce, recruit, and fill now take nearly a year.

I would be remiss not to relate what happens when the relationship between a concessioner and the NPS goes south. While there have been many attempts to introduce more competition into the field of national park concessions, the net result is that it is dominated by just a few powerful companies: Delaware North Companies Parks and Resorts, Aramark, Xanterra Parks and Resorts, Forever Resorts, with a few contracts going to Ortega National Parks, LLC, and Guest Services, Inc.

Individually and collectively, concessioners are a powerful lobby in Washington and work tirelessly to maintain their monopoly, compete among themselves, and keep NPS oversight in check. They do pay the NPS a franchise fee from their profits, but they also benefit from sections of the concessions law that not only protect their investments but allow them to grow over time. In one specific case, the value of the investments that the concessioner made into the facilities at the Grand Canyon had grown annually to that point that, on a monthly basis, the amount of money the NPS owed the company exceeded the amount the company paid the NPS in fees. This upside-down contract ultimately required the NPS to pay the company $100 million dollars.

Using the concept that competition for the opportunity to provide commercial services to the public will improve those services, I successfully opened up Mount Rainier guiding services to other companies beyond just the single concession, Rainier Mountaineering, Inc. Its founder and legendary climber Lou Whittaker threatened to have me fired.

As director, I oversaw the teams that analyze and issue "requests for proposals" (RFPs) for each concession contract at a particular park. These contracts, as Destry explained above, are unique to the National Park Service, issued for periods ranging from ten to twenty years, and can be quite lucrative to the winner. The Yosemite concession contract in particular is coveted as it grosses over $150 million a year, and the company can maintain nearly full occupancy with no advertising. As Destry noted, DNC, whose primary business is major ballparks, won the concession after it was purchased from Matsushita. DNC operated the concessions under renewals and extensions of the contract from 1993 until the NPS issued a new RFP in 2014. Besides DNC, Aramark also put in a bid. The NPS evaluated the contract proposals and found that Aramark offered the better proposal and would be offered the contract. In June 2014, in the middle of the process, I received a letter from DNC outlining the categories of assets for which they would require payment should the contract be awarded to another company. While such a list and their associated values is normal, what was not normal, and a complete surprise, was their listing of "intellectual property," valued at $51,200,000. Upon investigation, we learned that DNC had filed with the US Patent and Trademark Office and obtained the ownership of the names Yosemite, Ahwahnee, Wawona, Badger Pass, and many other place and facility names within Yosemite National Park.

I vividly remember sitting around my conference table with NPS senior staff, concessions specialists, and the DOI attorneys, and sharing our astonishment at this assertion by DNC. I will not repeat my language but I was furious that 1) a park concessioner would sneakily file for the legal intellectual property rights of the park name, 2) that I thought these names were in the public domain already, and 3) that the US Patent and Trademark people would even grant the right without calling us. It was clear to all of us that this was a ploy by DNC to retain the contract and force any competitor to back out rather than making a cash payment to DNC. While the contract is lucrative, an initial payout of $51 million was not viable. Attorneys from the Department of Justice were called in and again we discussed our options. I was sure not going to reward DNC with either the payment or a coerced renewal. The only option left was to change the names and go to court. So, with a great deal of angst, we changed the names of all the major concessions facilities in the park. The Ahwahnee Hotel became the Majestic Yosemite Hotel. The Wawona Hotel became Big Trees Lodge. Curry Village became Half Dome Village. And Badger Pass Ski Area was renamed Yosemite Ski and Snowboard Area.

We awarded the Yosemite contract to Aramark in June 2015, and the court case muddled along until a judge finally ordered a settlement in 2019. Aramark and the NPS paid DNC a total of $12 million and the names were restored. All concessions contracts now have a strict prohibition against the company filing for intellectual property rights on park names.

The US National Park Service and system is respected around the world and hosts millions of international visitors each year. Over 200 nations have established national park systems, and nearly all have requested technical support from NPS professionals. Managed out of a small office of International Affairs, NPS employees have traveled to parks in Costa Rica, South Africa, China, Cambodia, Italy, Australia, South Korea, and many other countries to assist in the planning, operation, and stewardship of their national parks. NPS leadership is called upon to participate and help lead major conferences such as the World Parks Congress, held every ten years. That said, during that last set of Republican administrations, the NPS was barred from this type of soft diplomacy. One conservative lobbyist told me he believed this type of international conservation work was all a conspiracy to lock up and prevent the mining of coal, oil, and gas in each nation. As the new NPS director in 2009, I traveled to the World Wilderness Congress being hosted in Mexico and was asked to speak. For the previous eight years, the NPS had been absent from this world effort to conserve protected areas. At the congress, there were thousands of representatives of parks and protected areas from around the world. I stepped to the microphone and said simply "We're back." The entire audience rose in applause.

Another government initiative that was used to undermine the mission of the NPS was GPRA: the Government Performance and Results Act of 1993. While there is nothing wrong with the agency setting goals, measuring results, and reporting on progress to the public, the implementation of the law by the political leadership of the Department of the Interior went awry. Again, using the private sector as the guide, the parks were required to prepare five-year plans on their activities against a set of measurable standards for each resource. If the resource in question already met the standard, then the park was not to expend any budget on that resource. Washington set the standards and, for example, the standard set for water quality was "swimmable." When this was applied to a park like Crater Lake, where the prime resource is the lake itself and the park was operating and funding a long-term research project on the lake's biology, under GPRA the park could not spend any budget on the lake research because the water was already "swimmable" (though quite cold!). The unique clarity of Crater Lake's water and its azure blue cast are qualities that need to be maintained above swimmable quality. Again, the NPS stalled and stalled and wrote many long and meaningless annual reports to satisfy the GPRA overlords in Washington. While the actual harm was minimal, it was an extraordinary waste of time and labor.

As a general rule, Republican administrations since 1972 have insisted that only the specific language of a park's enabling statute matters, while the broad overarching language of the Organic Act, which sets out the NPS mission, was irrelevant or could be ignored. Another iteration of this idea, dreamed up by the George W. Bush administration, was "core mission." In this case, each park was to establish, in writing, what its specific core mission was, based upon the exact language of the statute or proclamation creating the park. For instance, the "Act to preserve within Manassas National Battlefield Park, Virginia, the most important historic properties relating to the battle of Manassas, and for other purposes," approved April 17, 1954 (68 Stat. 56; 16 U.S.C. 429b)," clearly indicates the purpose of the park is to protect this Civil War battlefield. The law is silent regarding any other purpose, therefore, under core mission, any activity by the park that was not directly related to the preservation of the battlefield should not be done. On a field trip to Manassas National Battlefield with the political leadership of the department, I watched as the park superintendent revealed that they were hosting school field trips. I turned to one of the political appointees and asked if this was a core mission. The answer was no, education was outside of the core mission and should be stopped. This concept ignored the broad mandate articulated in the three laws that govern the entire system and used the narrow language of the specific park statute to interfere with the mission.

Actually, the idea that parks could be used as sources of public education, especially for children, has long been a target of the right-wing politicians who haunt the halls of DOI. Secretary Gale Norton issued orders banning park rangers from speaking in public school classrooms. Some of her supporters equated the NPS education programs for youth with creating little "brown shirts," referring to the indoctrinated Nazi youth of prewar Germany. When Secretary Norton was provided a copy of the NPS plan to engage youth in education programs ranging from the Civil War to geology, her counselor wrote in the margin of the report "mission creep." Most of the parks ignored her directive and figured out ways to engage youth in the parks through partners. When I became director, I was determined to get the NPS role in public education into statute so that no future Secretary could describe the NPS education programs as anything other than part of the mission. I succeeded with the National Park Service Centennial Act of 2016. Title III, section 301 states that the parks are "enhanced by the availability and use of a broad program of the highest quality interpretation and education."[7]

While most of the serious impacts to national parks come from the policy objectives of hostile administrations, occasionally Congress gets into the act through the appropriations process. At times, a rider on the appropriations bill will specifically prevent the NPS from implementing a specific policy or regulation. At other times, when Congress just fails to act on appropriations, it has a ripple effect across all government agencies, including the NPS. These periodic shutdowns of the US government are executed by the conservatives in the US Congress who refuse to pass an annual spending bill. In September of 2013, while I was serving as the director of the National Park Service, the Congress failed to pass either an appropriations bill or a continuing resolution, thereby ending all federal funding as of October 1. Considering that my primary responsibility was to protect the national parks "unimpaired for the enjoyment of future generations" and that leaving them open without staff would allow the public to do whatever they wanted there, I decided the best course of action was to close them all. Too often, the general public does not really understand the wide breadth of the federal government's role in our day-to-day lives. Surely many do not understand the impact of a government-wide funding shutdown until after it happens.

Lincoln Memorial, October 2013. Credit: National Park Service.

I issued a closure notice on October 1, 2013, shutting down the entire national park system, from Yellowstone to the National Mall. Occupants of park hotels and campgrounds were given three days to vacate. By October 3, the parks were emptied, gated, closed, and guarded by a small contingent of essential rangers who were ordered to work without pay. The remaining 21,328 employees were placed on unpaid furlough. The impact was almost immediate to gateway communities that rely on tourism, to the concessioners who provide lodging and food services to park visitors, and to park visitors who had saved for years for a planned visit to Yosemite or Yellowstone. Previous shutdowns had been during slow periods or off-season, but this one was in the fall, a peak season for visitation. Columbus Day weekend is extremely popular for travel, especially in the West.

A former NPS ranger and old friend of mine was now working for Utah's Grand County sheriff, in an area that includes popular parks like Arches and Canyonlands. He revealed to me that the sheriff was planning to forcibly open Arches National Park the following morning with his deputies and "armed vigilantes." The park was closed and gated and park rangers were guarding the gate. The scenario of a "blue on blue" (armed enforcement officers facing off) was untenable to me, the governor, and the Secretary. My team (made up of just a few who had not been furloughed) and I literally worked through the night to negotiate an agreement with the governor to transfer from the state of Utah to the NPS the daily cost to run Arches National Park. Secretary Sally Jewell signed off on the deal the following morning, money was wired to the Treasury, and we just barely averted a showdown. This set the precedent and, within a few days, other governors were transferring funds to the NPS to reopen parks such as the Grand Canyon.

The shutdown lasted until October 17, the longest in history up to that time.[8] While the situation was purely a political move by members of Congress, they were quick to blame me for all the bad press generated by the shutdown of the parks. On October 16, I was called before a joint committee hearing in the House of Representatives and grilled (without a break) for five hours on my decision to close the parks during the government shutdown. For me, the wholesale defunding of the National Park Service was the ultimate in interference in the mission, and I stood my ground in the hearing.[9]

My decision was vindicated when, in December of 2018, Congress once again shut down the federal government, but the Trump administration, wise to the public fallout from park closures and unconcerned about resource impacts, furloughed the park employees but left the parks open. The impacts were immediate and in some cases irreversible. Off-roaders plowed tracks into the fragile desert soils of Joshua Tree and even cut down the iconic trees. Across the system impacts included vandalism and break-ins of park buildings, cultural artifact hunting on historic battlefields, dogs off leash harassing wildlife, and people approaching wildlife or walking in dangerous areas. Leaving the parks open and unprotected was more than interference, it was intentional harm. Resource impairment, prohibited by the 1916 NPS Organic Act, occurred across the system due only to the Trump administration's interference with common sense and standard NPS policy.

So Long as Partisan Political, Rather Than Professional, Decision-making Rules the NPS, the Parks Are at Risk

As we have articulated in each chapter, the interference with the National Park Service's ability to achieve its legislative mission comes in all forms. Junior political appointees are given license to craft new policies or meddle in specific park affairs. Secretaries of the Interior propose draconian budget cuts, reorganization plans, or new privatization initiatives, and when they sense resistance, they reassign career employees to get them out of the way. Hostile members of Congress drag NPS leadership before hearings and question the very basis of park stewardship. Lobbyists for the concessioners work the halls of Interior and Congress to turn more and more of the park operations over to them for profit. NPS employees have a sad saying—"only the losses

Jonathan Jarvis at Yellowstone National Park, August 25, 2016. Credit: National Park Service.

are permanent"—which means the wins have to be fought again and again. While the NPS does win frequently, it should not be this hard to defend the very best of our nation's natural and cultural heritage.

This kind of repetitive undermining of the agency's mission has a profound effect on the morale of the career employees. The NPS consistently ranks low in the annual "Best Places to Work" survey. A close analysis of these results shows a high score in the match of employee skills to the mission and a low score for their ability achieve that mission.[10] As organizational consultant Margaret Wheatley, also a member of the National Park System Advisory Board, noted upon review of these results, this indicates a workforce that is "disappointed."

The time has come to curtail as much of this interference as possible, by getting the NPS out of the Department of the Interior.

SEVEN

INDEPENDENCE: FINDING A SUSTAINABLE FUTURE FOR A PERPETUITY AGENCY

DESTRY AND JONATHAN

Imagine for a moment that every four years there was a political purge of the top leadership of the Smithsonian Institution and a fresh group of political leaders arrived, with new ideas for this venerable and respected institution. Career museum directors would be shuffled around randomly to see which ones would retire. Scientists would be threatened and silenced if they wrote or spoke about anything the new leadership did not like or believe. Natural history museum exhibits that presented information about climate change, extinction, or evolution would be covered up or removed "for remodeling." At the African American Museum of History and Culture, the Civil Rights exhibits that show fire hoses and attack dogs being unleashed on peaceful citizens would now declare that there "were good people on both sides." Plans would be formulated and aired to turn the museums over to the private sector, and business executives from amusement parks would be invited to suggest ways to monetize the institution including entrance fees, private rentals, selling of artifacts, and naming rights. And at the Air and Space Museum, new exhibits would present "alternative facts" about the 1969 moon landing.

Sounds horrifying? This is the situation that faces the National Park Service every four to eight years. In our nearly five decades we have heard and seen it all. We have personally witnessed constant proposals—not just hearsay—to turn the parks over to the private sector, ban all references to evolution, include suggestions that the Grand Canyon is only a few thousand years old, increase fees without public involvement or review, and install new park exhibits presenting the current president's priorities or, worse, carving Donald Trump's face onto Mount Rushmore. It was bad enough for Trump to use Mount Rushmore for a reelection campaign rally, or for Vice President Pence to do the same at Fort McHenry. The national parks are places to come together as a nation, not for partisan politics.

We have watched while senior career professionals, including scientists, were forcibly removed or reassigned, climate change exhibits removed, and scientific reports altered. We have had to explain, repeatedly, why it would be a bad idea to open all the national parks to sport hunting. We saw scholarly research projects on African American history stopped. And we have resisted the periodic idea that Ronald Reagan's image be carved into the rock face of El Capitan at Yosemite National Park.

During the administration of Donald J. Trump, the presidentially appointed, Senate-confirmed position of NPS director went unfilled for the entire four years. Every Senior Executive Service park superintendent and regional director was forcibly moved, causing many to retire. National Park Service websites on climate change went down "for updating" and interpretive exhibits were removed. Park superintendents were prohibited from talking to the media about park impacts from extractive activities adjacent to park boundaries. Parks were left open to the public but unstaffed during the government shutdown and the COVID-19 pandemic, with much resultant vandalism and damage to park resources from uncaring members of the public. Every budget from the administration proposed draconian cuts to park operations. While one might suggest that this is just the way it is all supposed to work with a change in administration, these changes actually impact the overarching mission of the NPS as stated in law: "to preserve the national parks unimpaired for the enjoyment of future generations."

Every four to eight years, we hand over the keys to our parks and public lands to political appointees at the Department of the Interior who view the world through a commodity lens. When Jonathan was inducted into the Senior Executive Service in 2002, he attended a presentation by the incoming George W. Bush administration's political leadership for the Department of the Interior. One presenter said that Sequoia National Park, as an example, must earn enough fees at the entrance gate to operate, just like Kings Dominion amusement park in Virginia does; otherwise, the national park "has no right to exist."

Ours are not the first voices, and surely will not be the last, that advocate for moving the National Park Service out of the highly conflicted Department of the Interior. We believe there is ample precedent, and that a strong case can be made for doing so. Clearly, some DOI Secretaries have fully embraced the NPS mission, while others surely have not. The debilitating effects of the whipsaw of political appointees have prevented this agency, tasked with preserving America's best natural and cultural places in perpetuity, from being the best that it could be. The American people deserve better management, preservation, and interpretation of every one of the places of national significance that have been set aside for their enjoyment, which, based on the past fifty years of observation and analysis, cannot be achieved without a major change—a change that independence would provide.

When former NPS Director George Hartzog made his wry observation that "if you show me your budget, I will tell you your policy," he likely did not anticipate how quickly his politically appointed DOI bosses would use budget cuts, fostered by Nixon's Office of Management and Budget, and a political litmus test, to control the NPS and to punish its career professionals. Every DOI Secretary in a Republican administration since 1972 has done so. What was reality inside the NPS became abundantly clear to the public when James Watt opined that he would change either the policy or the person to get what he wanted.

In 1977 President Carter proposed a Department of the Environment that included the Environmental Protection Agency, the NPS, the US Fish and Wildlife Service, and the National Marine and Estuarine Sanctuaries Programs from NOAA. That proposed movement of DOI agencies would have brought together all of the federal agencies with conservation and environmental protection missions. The NPS and the USFWS have far more in common with each other than does their current alignment with agencies that primarily focus on natural resource development and extraction. That reorganization did not gain sufficient support in Congress.

One other such realignment proposed in Congress did focus on the NPS specifically. Representative Bruce Vento was chair of the House National Parks Subcommittee in 1982 when he convened an oversight hearing on the plight of the NPS. Vento, a science teacher before being elected to Congress, well recognized the situation of the NPS and its inability to truly conserve the parks unimpaired while under the political control of Secretaries of the Interior like James Watt. Chair Vento convened his first hearing of this subcommittee in June 1982. The lead witness was former NPS Director Bill Whalen (1977–1980), who testified that:

> I have come to a conclusion that has been formulating in my mind for the past five years . . . that the National Park Service should be removed from the jurisdiction of the Department of the Interior. . . . The National Park Service, like the Smithsonian, has a principal mission which is to preserve and

protect the treasures of America and to interpret to the public not only what America was, but what America is, and what America might be in the future. There are many reasons why the National Park Service should be removed from the Department of the Interior and they center primarily around the fact that the Interior Department is a "confederation of bureaus" with often conflicting views. When decisions must be made by the Department on land that surrounds or is near a national park unit, many times the views of the professional resource manager of the National Park Service are ignored or watered down. . . . The National Park System, which is being heavily impacted by sources external to its boundaries, should have the ability to clearly express its unexpurgated views when external impacts are threatening its pristine existence.

Throughout 1982 and early 1983, the committee continued to hold hearings on the threats to the national parks. In March 1983, Representative John Seiberling and eighty-five bipartisan cosponsors introduced HR 2379, the National Park System Protection and Resources Management Act. While this bill stopped short of making NPS an independent agency, it did seek to greatly constrain the discretion of the DOI Secretary regarding actions of other DOI agencies. Section 10(b) of that bill stated, "In any case of areas which are adjacent to any unit of the national park system, where the Secretary is vested with any authority . . . to issue any lease . . . authorize or permit any use, . . . sell or otherwise dispose of such lands . . . the Secretary may exercise such authority only after he has determined that the exercise of such authority will not have a significant adverse effect on the values for which such national park unit was established."

For agencies outside of the DOI, the bill went on to provide that "Each agency or instrumentality of the United States conducting or supporting activities within or adjacent to any unit of the national park system shall, to the extent practicable, undertake to insure that those activities will not significantly degrade the natural or cultural resources or values for which the unit was established."

Both provisions of the bill sought to achieve federal consistency to protect park resources and values. In May, the bill was favorably reported from the committee to the full House, but enactment was opposed by the Reagan administration, and the bill died without Senate action.

In February 1988, Chair Vento introduced HR 3964, another move toward independence from the DOI. That bill would have established an independent National Park Review Board, required the NPS director to be confirmed by the Senate, and to be qualified "by training and experience and by demonstrated ability to administer, protect and preserve the natural and cultural resources of the United States." While leaving the NPS located within the DOI, the bill further provided that "all functions and authorities of the Secretary which are carried out through the National Park Service as of July 1, 1988, shall be transferred to and vested in the Director of the National Park Service.... The Director and his officers and employees of the National Park Service shall not be responsible to, nor subject to the supervision or direction of any officer, or employee, or agent of any part of the Department of the Interior." The bill passed the House in July 1988 but failed in the Senate.

At that time, nearing the end of President Reagan's second term, the NPS had been fighting with DOI political appointees on hundreds of issues, reversals, and constraints to its policy, budget, and personnel decisions. The political structure of the DOI then, and through at least 2020, imposes at least nine layers of politically appointed decision makers between the NPS director and the Secretary. Not every one of these political appointees interacts with the NPS on every issue, but all can, and often do, usually to impose a political interpretation on a policy matter that is better left to professional judgment based on science and long experience in park management.

Clearly, the NPS has become a political football, with wide pendulum swings in funding and policy from one administration and Congress to the next. In yet another example, Republican administrations' DOI and OMB political appointees have regularly zeroed out NPS budget requests for funding the Urban Parks and Recreation Recovery program, a matched grant program for city park systems, "because those cities are all run by Democrats."

Such abrupt policy shifts will not stop so long as the NPS is one agency among many others with conflicting missions all rolled up into the Department of the Interior. The same Secretary must manage mineral leasing for coal, oil, and gas on public lands and on the outer continental shelf, regulate lands for hunting and fishing, sustain hard rock mining, care for Native American Trust lands and tribal economies, make deserts bloom with scarce water sources, and conserve the national parks unimpaired.

Truly, the much-discussed perception of a "dual mission" is not within the NPS but with the Secretary of the Interior and the political appointees below him/her. Inevitably, DOI Secretaries come into that job focused either on maximum resource development—of BLM lands, Tribal Trust lands, outer continental shelf lands, or Bureau of Reclamation water resources (Watt, Hodel, Luján, Norton, Kempthorne, Zinke, and Bernhardt)—or they come in seeking to reform those functions (Udall, Andrus, Babbitt, Salazar).

No Secretary in the last fifty years has made solving the problem of conserving the national parks in perpetuity a top priority—there is just too much else to be done. No Secretary has been willing to treat the NPS as anything other than just another agency under his/her control.

For its first half-century, the NPS was regarded as a nonpartisan, professionally led and staffed agency, but, in many of the administrations in its second half-century, the NPS position on most matters has not been determined by the professional views of the agency but by the political appointees around it. These politicians force the NPS to be for or against a new area, policy, or issue based on who proposed it, which party they were from, and how the matter relates to other concerns of the administration, especially the next election.

Instead of the NPS remaining the overworked and under-appreciated Cinderella of the DOI, we see a better, more independent future for the agency as a necessity if these places of national significance are to be conserved unimpaired. But this can only happen by making the National Park Service an independent agency.

Opponents of this idea argue that the agency needs the power and prestige of a cabinet secretary in order to compete effectively for attention, funding, staffing, and policy support. They believe that the NPS would not fare well having to go it alone. Quite to the contrary, the NPS already has more prestige than any DOI Secretary, and more public support and recognition than any other public land agency due to the power of the parks themselves. The 2016 Centennial of the National Park Service again demonstrated their broad and deep support. That year, the parks hosted more visitors than the National Football League, National Basketball Association, Major League Baseball, US Soccer, and the National Association of Stock Car Racing combined. In addition, according to the study conducted and published by Harvard economist Linda Bilmes and Colorado State University economist John Loomis, the value of the national park system is over 100 billion dollars, and the public is willing to increase its support of the parks through increased taxes.[1] What the NPS lacks at present is the ability to express its own professional positions, advocate for its own budget, and defend the parks from the threats imposed by the proposed actions of other agencies.

Freeing the NPS from the Department of the Interior does not free the agency from the three primary powers of Congress: oversight, legislation, and budget appropriation. That said, the US Congress conducts its responsibilities in the public view, through hearings and testimony, and that transparency helps reveal and curtail the worst of the legislative ideas. In contrast, a hostile group of political appointees in the Department of the Interior can carry out their policy objectives in complete darkness, executing them long before the public is aware. That the Congress is there to provide oversight of an independent NPS is fully appropriate, as it provides checks and balances to independence.

There is no exact model for what an independent NPS would look like or exactly how it would be structured. However, elements that already exist within the frameworks of the Environmental Protection Agency, the National Archives and Records Administration, and the Smithsonian Institution point the way. Legislation would be required when the time is right. These are our suggestions for key elements of an independent National Park Service.

LEADERSHIP

The independent National Park Service could be governed by a Board of Regents, similar to the Smithsonian,[2] with equally balanced membership representing the three branches of government: executive, legislative, and judicial. The board would also have an equal number of members from the public who represent the diversity and geography of the nation.

The Board of Regents would nominate the director of the National Park Service, who would be confirmed by the Senate and serve for one fixed term of six years. The individual, who could be renominated, would be required to meet a set of professional qualifications and experience in protected area conservation management, natural and social sciences or history, and public service leadership.

The director would be directly responsible for the selection or the delegation of selection of all senior positions in the agency, according to normal federal hiring policy and regulations.

The National Park System Advisory Board charter would be expanded to include more active public oversight of the operations, expansion, and function of the National Park Service. The members of the advisory board would be selected by the Board of Regents in coordination with the NPS director, while retaining the requirements of current law for the relevant qualifications of board members.

FEDERAL FUNDING

The National Park Service would formulate and submit its federal budgetary request to Congress on a two-year basis rather than annually. A two-year budget for the National Park Service would give it greater flexibility in its highly seasonal operations. The NPS would remain eligible for funding from other federal appropriations as it is currently, including the Land and Water

Conservation Fund, Federal Highways Transportation funds, fire suppression funds, and so on.

NONFEDERAL FUNDING

Fees. Currently the NPS collects over $300 million annually in fees, which are retained by the US Treasury in noninterest bearing accounts for return to the agency. The independent National Park Service would be authorized to collect and retain all entrance, campground, and other special use fees and invest all or a portion in interest-bearing accounts to be managed by the National Park Foundation. Any new fees or increases in fees would be subject to public review and have to be approved by the Board of Regents.

Philanthropy. The National Park Foundation would be the primary philanthropic partner of the National Park Service. Board members would be nominated by the Board of Regents and selected for their proven track record in philanthropic fund-raising in the fields of conservation and historic preservation. Currently the NPF raises approximately $100 million per year. The NPF would work cooperatively with the individual park Friends organizations.

Visitor Service Operations. The hotel, food, beverage, guide, and outfitting private sector operations within the National Park Service are under the Concessions Management Act of 1998 and generate over a billion dollars of revenue per year. The independent National Park Service would invoke the new authority in the National Park Service Centennial Act (section 701)[3] and transition out the concessioners as their contracts expire. The new "business side" of the National Park Service would then offer fee-for-service contracts to operate and maintain the commercial facilities in the parks rather than standard concession contracts. Under this model, a commercial company would operate a park hotel under a contract for a set time and fee, with the profits from the operation going to the NPS. This could generate hundreds of millions for the NPS, much more than do the current franchise fees paid to the NPS for the privilege of operating in a national park.

PARTNERSHIPS

The independent National Park Service would be broadly authorized to enter into cooperative agreements, including a major expansion of the use of conservation and service corps, like the national Student Conservation Association or state-based groups like the Montana Conservation Corps. Currently, service and conservation corps provide a few thousand work crews and interns annually to carry out needed park work. This number could easily grow to many thousands more, with funds established for that purpose.

COMANAGEMENT WITH NATIVE AMERICAN TRIBES

As previously described, the history of the national parks is one of removal, disenfranchisement, and regulatory restrictions for the Native Americans whose traditional territories encompassed park lands. There have been some attempts to reconcile this history with new agreements, invitations for tribal members to tell their own stories to the public, and the finalization of the Native American plant collecting regulation. But much more can be done to engage tribes in using their traditional ecological knowledge, protecting archeological and sacred sites, and continuing traditional practices in ways that are compatible with the NPS mission. And NPS independence should include new authority for comanagement of park lands with traditionally associated Native American tribal governments.

As an independent agency, the National Park Service and system would:

- be composed of all of the best and most representative natural and cultural sites of national significance in America, including sites of shame, sites of conscience, and those that fully represent all of the cultures found in America.
- annually submit a list of potential new parks that would meet the objectives outlined above;
- have completed a full biological and cultural resources inventory of what each of these best places contains;
- have extensive monitoring systems in place to identify and understand changes in these best places and, through applied research and resources management, be able to implement decisions that assure perpetuation of the system "unimpaired";
- have sufficient educated and well-trained personnel to manage these diverse resources;
- conduct extensive public education programs reaching all ages and ethnicities in America in a manner that shows the relevance of the system for every American, and that builds understanding and constituency;
- have true and meaningful partnerships in place that complement the central functions of the service and add a true measure of excellence to management of the system;
- seek comanagement agreements with every federally recognized Native American tribe whose ancestral lands lie within park boundaries;
- have significant and sustaining political champions and the public support that assure the parks receive necessary funding and adopt appropriate policies;

- have new authorities of its own, or new laws governing other agencies, so that adjacent land use practices are fully compatible with the mission of the service and preservation of the system;
- adopt fully sustainable practices, including greening of facilities and visitor transportation systems.
- have control over the business side of the parks—fees, tourism, concessions, and gateway community relations so that these important functions complement, rather than conflict, with the core preservation mission;
- administer an array of conservation, recreation, and historic preservation support programs, including grants and technical assistance, in support of the work of other park agencies and organizations at the state and local levels;
- be fully funded and efficiently administered, with significantly reduced layers of political control and uninformed bureaucracy; and
- base its decisions upon the best available science, fidelity to the law, and the long-term public interest.

Over the more than ninety years of our combined careers, we have been in the arena, in the words of President Teddy Roosevelt, "marred by dust and sweat and blood," and have learned to dust ourselves off and get back into the fight. We have learned that those who would view the national parks as just another resource to exploit will turn to the politics of personal destruction if all else fails. We have learned that when you take political people into a national park, almost every time, their attitudes change, they become more open and willing to support park stewardship, and, with this in mind, we have hosted countless trips to parks with the underlying goal of conversion. And we have learned never to give up on the mission of the National Park Service.

We have spent our entire adult lives fighting for the protection of our national parks. When we get together with our families of children and grandchildren, we are often off in the corner, debating the latest environmental transgression, park legislation, departmental political appointee, or threat to our national parks. Our wonderful spouses Barb and Paula are used to it and give us space until we run out of steam.

We two brothers don't agree on everything, but, whether looking at the National Park Service from the inside or the outside, we agree on all the essentials of the agency's mission and the necessity to protect our parklands for future generations. We learned from our family the value of nature and of history, and from all with whom we have worked, the value of the parks in protecting natural and cultural resources. We know we are never alone in this fight, as millions of people enjoy the benefits of their national parks. The 320 million people who visited national parks last year were not all of one political persuasion, one ethnicity, or one demographic. The National Parks are a place where we can come together to agree, again in the words of Teddy Roosevelt, that they must be "preserved for their children and their children's children forever, with their majestic beauty all unmarred." It will be those children who will judge our actions as to whether we truly did preserve the national parks, unimpaired, for their enjoyment and inspiration.

NOTES

Foreword

1   Yosemite Valley Grant Act, Senate Bill 203, June 30, 1864.
2   In 1832, Hot Springs, Arkansas, had been established as a "reservation." It was later designated a "public park" in 1880, after both Yosemite and Yellowstone became "national parks."
3   Wallace Stegner, quote from Ken Burns and Dayton Duncan, *The National Parks: America's Best Idea* (New York: Alfred A. Knopf, 2009).

Introduction

1   National Park System Advisory Board Report, *Rethinking the National Parks for the 21st Century* (National Geographic Society, Washington, DC, 2001).
2   54 US Code, Chapter 1001.
3   "Outdoor Recreation for America: A Report to the President and to the Congress by the Outdoor Recreation Resources Review Commission," Library of Congress Catalog No. 62-60017 (Washington, DC: Government Printing Office, January 1962).
4   "One-Third of the Nation's Land: A Report to the President and to the Congress by the Public Land Law Review Commission" (Washington, DC: Government Printing Office, June 20, 1970).
5   Report of the Secretary's Advisory Board on National Parks, Historic Sites, Buildings and Monuments, June 1970 (D. Jarvis, personal copy).
6   The formal name varies between administrations—Fish, Wildlife and Parks, Fish and Wildlife and Parks, but refers to the same Assistant Secretary.

Chapter 1

1   "The Recreation Imperative," Committee Print published at the request of Henry M. Jackson, Chairman, Committee on Interior and Insular Affairs, US Senate, September 1974.

2   House of Representatives Report 94-1318, *The Degradation of Our National Parks*, June 30, 1976.

3   "Phil Burton was the most naturally gifted elected official or politician I have ever known or run across. All of his habits were tailor-made for politics. He had an appetite for detail beyond belief on every issue. He had an unlimited amount of energy. He had supreme confidence, and he was absolutely devoid of a need to be loved. I think Phil Burton believed that he could absolutely make a difference in any situation, and he usually did, and that drove him more than anything else."—California Assembly Speaker Willie Brown, quoted in John Jacobs, *A Rage for Justice: The Passion and Politics of Phillip Burton* (Berkeley: University of California Press, 1995), p. xix.

4   Today, the district is represented by House Speaker Nancy Pelosi. Ironically, the year before Wayburn's visit (1976), Burton had lost his race to be Speaker of the House by one vote, and Rep. Leo Ryan was the only Democrat in the California delegation who voted against him. The two were not friends but still worked together to save the Redwoods.

5   House Report 95-106, *Protecting Redwood National Park*, Report of the Committee on Government Operations, March 23, 1977.

6   Redwood Expansion Act of 1978, P.L. 95-250, 92 Stat. 163, March 27, 1978.

7   National Parks and Recreation Act of 1978. P.L. 95-625, 92 Stat. 3467, November 10, 1978.

8   The booklet was paid for entirely with donated funds.

9   Rep. Jim Hansen, quoted in the *Denver Post*, November 6, 1994.

10   Printed statement of NPS Director Roger Kennedy, in testimony before the Subcommittee on National Parks, Forests and Lands, February 23, 1995.

11   If the actions sound familiar, it's because most of these policies, and many of the same political appointees, came back to the DOI during the Trump Administration.

12   As I hoped, Congress did respond within a few years. The Great American Outdoors Act, Public Law No. 116-152, passed in August of 2020, significantly increasing the funding for the NPS.

13   The Second Century Commission was an independent panel of experts brought together by the National Parks Conservation Association to prepare a report to guide the NPS into its next hundred years; https://www.npca.org/resources/1900-national-parks-second-century-commission-report.

14   These reports built on past theme studies on the contributions of African Americans. They include *Asian American Pacific Islander National Historic Landmark Theme Study*, National Park Service, 2017; *American Latinos and the Making of the United States: A Theme Study*, National Park System Advisory Board, 2013; *LGBTQ America: A Theme Study of Lesbian, Gay, Bisexual, Transgender, and Queer History*, National Park Foundation, 2016. All three can be found at https://www.nps.gov/subjects/tellingallamericansstories/index.htm.

15   National Trust for Historic Preservation: https://savingplaces.org/stories/the-forgotten-the-contraband-of-america-and-the-road-to-freedom#.

16   President Barack Obama, *Remarks by the President at the Dedication of the César E. Chávez National Monument*, National Archives, October 8, 2012.

17   See https://parkplanning.nps.gov/files/NationalParkServiceSystemPlan2017.pdf.

Chapter 2

1   George Hartzog, *Battling for the National Parks* (New York: Moyer Bell Limited, 1993), chap. 16.

2   See chap. 1 for more on this.

3   Our coalition slogan for the campaign became "The Last Great Chance To Do It Right the First Time." It refers to the fact that most earlier national park campaigns had resulted in compromised boundaries, necessitating repeated subsequent attempts to increase the acreage of the park to include whole watersheds or complete ecosystems, rarely with success.

4   The Brown family owned Brown Distilleries, makers of Jack Daniels and other whiskeys, Herradura and other tequilas, Glenglassaugh and other Scotch, Finlandia Vodka, and various other wines, liqueurs, and gin. In the mid-1980s, I had the pleasure of escorting Sally to Valdez, Alaska, the place of her birth in 1911. Her father, an army lieutenant, was in charge of building the first road and telegraph line to Fairbanks. She had not been back since her youth.

5   In late January 2021, Trump's DOI Secretary Bernhardt approved a right-of-way through Gates of the Arctic National Park for this 200-mile road which will cross eleven rivers and over a thousand smaller streams.

6   Malcomb Roberts, *The Wit and Wisdom of Wally Hickel* (Anchorage: Searchers Press, 1994).

7   John Kauffmann, *Alaska's Brooks Range: The Ultimate Mountains* (Seattle: Mountaineers, 1988), p. 169.

Chapter 3

1   Letter from NPS Director Bill Whalen to Destry Jarvis, NPCA, October 7, 1977 (D. Jarvis, personal copy).

2   National Park System General Authorities Act, P.L. 91-383, 84 Stat. 825, August 18, 1970.

3   Redwood Expansion Act of 1978, P.L. 95-250, 92 Stat. 163, March 27, 1978.

4   Typed transcript, Address of James G. Watt, Secretary of the Interior, Conference of National Park Concessioners, International Hotel, Washington, DC, March 9, 1981 (D. Jarvis, personal copy).

5   NPS Director Bill Mott, memo to regional directors, February 6, 1986, transmitting December 1985 Annex #2 policy decision memo from Assistant Secretary Horn (D. Jarvis, personal copy).

6   NPS Director Bill Mott, memo to regional directors, February 6, 1986.

7   NPS Director Bill Mott, memo to regional directors, February 6, 1986.

8   Assistant Secretary Horn, memo to NPS Director Mott, February 18, 1986 (D. Jarvis, personal copy).

9   NPS Deputy Director Denis Galvin, memo to Assistant Secretary Horn, February 24, 1986 (D. Jarvis, personal copy).

10   NPCA News Release, March 6, 1986, p. 2 (D. Jarvis, personal copy).

11   NPS Deputy Director Denis Galvin, memo to Assistant Secretary Horn, May 15, 1986 (D. Jarvis, personal copy).

12   Assistant Secretary Horn, memo to NPS Director Mott, June 6, 1986 (D. Jarvis, personal copy).

13   NPS Director Mott, memo to Assistant Secretary Horn, July 15, 1986 (D. Jarvis, personal copy).

14   NPS Director Mott, memo to Assistant Secretary Horn, September 11, 1986 (D. Jarvis, personal copy).

15   NPS Director Mott, memo to Assistant Secretary Horn transmitting a rewrite of chap. 1 of NPS *Management Policies*, September 30, 1986 (D. Jarvis, personal copy).

16   Assistant Secretary Horn, memo to NPS Director Mott, September 30, 1986, p. 1 (D. Jarvis, personal copy).

17   Assistant Secretary Horn, memo to NPS Director Mott, November 25, 1986, p. 1 (D. Jarvis, personal copy).

18   Assistant Secretary Horn, memo to NPS Director Mott, November 25, 1986, p. 2.

19   Phil Shabecoff was the best and most focused environmental reporter in the country at that time. He had a thirty-two-year career at the *New York Times*, the last fourteen as their chief environment reporter. After he left the *Times*, he founded and was publisher of an online environmental news site, *Greenwire*. His book, *A Fierce Green Fire: The History of the Environmental Movement*, was published by Hill and Wang in 1993.

20   NPCA letter to NPS Director Mott, June 9, 1987 (D. Jarvis, personal copy).

21   Letter from NPS Associate Director Davis to D. Jarvis, June 26, 1987 (D. Jarvis, personal copy).

22   NPCA written comments on proposed revisions to NPS Management Policies, May 27, 1988 (D. Jarvis, personal copy).

23   NPCA News Release, May 27, 1988 (D. Jarvis, personal copy).

24   Chairman Bruce Vento, Subcommittee on National Parks, letter to NPS Director Mott, June 10, 1988 (D. Jarvis, personal copy).

25   Mammoth Cave National Park Superintendent Dave Mihalic, memo to NPS Director Mott, June 20, 1988 (D. Jarvis, personal copy).

26   Everglades National Park Superintendent Rob Arnberger, memo to NPS Director Mott, May 6, 1988 (D. Jarvis, personal copy).

27   For the NPS, at the end of a nearly four-year process, "Reinventing Government" meant moving more headquarters staff to the field and reducing regional offices from ten to seven, eliminating Seattle, Santa Fe, and Boston as formal regional offices but not fully closing them.

28   Assistant Secretary Don Barry, memo to NPS Director Bob Stanton, August 6, 1999, p. 2 (D. Jarvis, personal copy).

29   Assistant Secretary Don Barry, memo to NPS Director Bob Stanton, August 6, 1999, p. 2.

30   Assistant Secretary Don Barry, memo to NPS Director Bob Stanton, August 6, 1999, p. 2.

31   NPS Director Bob Stanton, memo to DOI Secretary Bruce Babbitt, signed by Babbitt on January 7, 2000 (D. Jarvis, personal copy).

32   US District Court of the District of Utah, Central Division Case No.2:95CV559DAK 387 F. Supp.2d 1178:2005 U.S. Dist. LEXIS 26365.

33   US District Court of the District of Utah, Central Division Case No.2:95CV559DAK 387 F. Supp.2d 1178:2005 U.S. Dist. LEXIS 26365.

34   Memo to the Deputy Director, August 5, 2005 (J. Jarvis personal copy).

35   Ken Salazar would become Secretary of the Interior in 2009 and recommend Jonathan to President Obama to serve as director of the NPS.

36   Grinnell Resurvey: https://mvz.berkeley.edu/Grinnell/.

37 National Park System Advisory Board, Revisiting Leopold: A Report of the National Park System Advisory Board, 2012, p. 11; https://www.nps.gov/calltoaction/PDF /LeopoldReport_2012.pdf.

38 National Park System Advisory Board, Revisiting Leopold, p. 15.

39 Tony Knowles, https://www.washingtonpost.com/national/health-science/nearly-all -members-of-national-park-service-advisory-panel-resign-in-frustration/2018/01/16 /b322ef5e-fae3-11e7-ad8c-ecbb62019393_story.html.

Chapter 4

1   George Melendez Wright, *Fauna Series No. 1* (National Park Service, 1932).

2   The Bureau of Biological Survey was transferred from USDA to DOI in 1939 and evolved into the present-day US Fish and Wildlife Service.

3   The *Fauna Series* began in 1932 with *Preliminary Survey of Faunal Relations in National Parks* and was followed in 1933 by *Wildlife Management in National Parks.*

4   Senator Lee Metcalf, quoted in Senate Committee on Appropriations Hearing Record, 1968, p. 2451; https://www.google.com/books/edition/Hearings/DsdTluM3mD4C?hl =en&gbpv=1.

5   "Wildlife Management in the National Parks," Report of the Advisory Board on Wildlife Management appointed by Secretary of the Interior Udall, March 4, 1963 (D. Jarvis, personal copy).

6   "A Report by the Advisory Committee to the National Park Service on Research," National Research Council, National Academy of Sciences, August 1, 1963, p. 2 (D. Jarvis, personal copy).

7   The Federal Office of Personnel Management, the agency that classifies every federal job, originally classified the Park Ranger series (025) as "blue collar" jobs, which had no positive education requirement. In 1994, the Park Ranger jobs were reclassified to require a positive education requirement. The park ranger law enforcement jobs have detailed training requirements, the same as any other federal law enforcement personnel.

8   Bruce M. Kilgore, "Views on Natural Science and Resource Management in the Western Region," keynote address at the NPS Pacific Northwest Region, Science/Resources Management Workshop, April 18–20, 1978 (D. Jarvis, personal copy).

9   Printed statement of NPS Director Mott testifying before the Senate Committee on Public Lands, July 18, 1985, p. 3 (D. Jarvis, personal copy).

10 Destry Jarvis speech, NPS Fourth Conference on Science in the National Parks, July 13–18, 1986, sponsored by the George Wright Society (D. Jarvis, personal copy).

11 Horn had spent his earlier political career working for Alaska Rep. Don Young, opposing enactment of the Alaska Lands Act; see chap. 2.

12 The Senior Executive Service was authorized by a civil service law in the 1970s; the SES offers a higher pay grade for career civil service but also allows higher-level political appointees to transfer these employees to another position or another department against their wishes, or even to force them to retire.

13 Howard Chapman, quoted in "Undermining the Park Service," *Los Angeles Times*, December 29, 1986. In his effort to avoid being forced to retire, Chapman, a highly effective NPS career manager, chose to go public with his objections, testifying before Congress and engaging in several media interviews, including with the *New York Times*, the *Los Angeles Times*, and the *Fresno Bee*.

14 "Concluding Recommendations of the NPS Director's Task Force on Biological Diversity," Report to the Director of the National Park Service on the Role of the National Park Service in Protecting Biological Diversity, March 1987, p. 51.

15 Dave Simon was a Yale History graduate under Dr. Robin Winks, himself one of the most distinguished park experts and advocates and an NPCA board member. Dave later moved on to be NPCA's Southwest regional director, then was New Mexico State Parks director under Governor Bill Richardson, and is currently City Parks director for Albuquerque, New Mexico.

16 "National Parks: From Vignettes to a Global View," Report from the Commission on Research and Resource Management Policy in the National Park System, March 19, 1989.

17 "National Parks for the 21st Century: The Vail Agenda," Report and Recommendations to the Director of the National Park Service from the Steering Committee of the 75th Anniversary Symposium, March 25, 1992.

18 Dr. Paul Risser, "Science and the National Parks," National Research Council (Washington, DC: National Academy Press, 1992).

19 Robert Cahn, review of "Science and the National Parks," *Environment* 35, no. 2 (March 1993).

20 Snowmobiles are allowed on some of the unpaved carriage paths in Acadia National Park, for example.

21 Winter Use Plan, Yellowstone National Park, Environmental Impact Statement, 1999.

22 "Weird Science—The Interior Department's Manipulation of Science for Political Purposes," Committee on Resources, U.S. House of Representatives, December 17, 2002.

23 "Weird Science," p. 7.

24 Statement of Michael Finley, Coalition to Protect America's National Parks, National Press Club, 2003 (D. Jarvis, personal copy).

25 Roland Wauer, *My Wild Life* (Lubbock: Texas Tech University Press, 2014), pp. 146, 174.

26 As a side note, the rangers began to shift much more strongly toward law enforcement, especially with the legislative action that granted them the twenty-year retirement (known as 6c) shared by other law enforcement agencies and firefighters. This also carried mandatory retirement at fifty-seven, which resulted in fewer and fewer rangers aspiring to become superintendents. This void was then filled by other professionals, especially resource managers.

27 I learned in the middle of the negotiations that Volker had worked as a seasonal ranger in North Cascades National Park, specifically in the Stehekin district, and had detailed knowledge of the park.

28 Richard Sellars, *Preserving Nature in the National Parks* (New Haven, CT: Yale University Press, 1997).

29 I confronted the deputy director of the NPS at the time, Deny Galvin, about why, after all the years of lobbying by the resource management specialists, he had suddenly had a deathbed conversion to support the Challenge. He wryly responded "because you were all so inarticulate."

30 Summer Brennan, *The Oyster War: The True Story of a Small Farm, Big Politics, and the Future of Wilderness in America* (Berkeley, CA: Counterpoint Publisher, 2015).

31 Secretary Salazar told me afterward that he knew that I would resign if he made the

decision to keep the oyster farm—but that he would have fired me first so that my threat to resign over a policy decision would not be a factor in his decision making.

32   National Park Service, Director's Order 100, December 20, 2016; rescinded August 16, 2017, and not yet reinstated; https://www.nps.gov/policy/dorders/do_100.htm.

33   Written testimony of Dr. Maria Caffrey before the House Committee on Natural Resources, July 25, 2019, Washington, DC; https://naturalresources.house.gov/imo /media/doc/Dr.%20Caffrey%20-%20Written%20Testimony%20-%20FC%20Ov %20Hrg%2007.25.19%20(Scientific%20Integrity).pdf.

34   The Restoration Project: https://rproject.world/.

Chapter 5

1   National Park Service Centennial Act, Public Law 114-289; section 401 relates to the National Park Foundation.

2   Hearing before the House Subcommittee on Civil Service, The Directed Reassignments of John Mumma and L. Lorraine Mintzmyer, September 24, 1991.

3   John Freemuth, *Islands under Siege: National Parks and the Politics of External Threats* (Lawrence: University Press of Kansas, 1991).

4   Title 36 Code of Federal Regulations, part 2.

5   Department of the Interior, Memorandum, Options Regarding Applications for Hardrock Mineral Prospecting Permits on Acquired Lands Near a Unit of the National Park System, April 16, 1998; https://doi.opengov.ibmcloud.com/sites/doi.opengov .ibmcloud.com/files/uploads/M-36993.pdf.

6   National Park System Advisory Board, Revisiting Leopold.

Chapter 6

1   National Park Service Organic Act (16 USC 1), August 25, 1916.

2   National Park System General Authorities Act, P.L. 91-383, 84 Stat. 825, August 18, 1970.

3   Redwood Expansion Act of 1978, P.L. 95-250, 92 Stat. 163, March 27, 1978.

4   This is a direct quote from the assistant administrator of the FAA to me and other NPS representatives in 1994 as we attempted to reconcile our conflicting missions.

5   Lynn Scarlett is one of the few conservative political appointees within the Department of the Interior to undergo a conversion to greater support for conservation during her tenure. I witnessed a profound change in Scarlett as she became more aware of the dedicated professionals working hard for conservation.

6   Michael Shnayerson, "Who's Ruining Our National Parks?": https://www.vanityfair .com/news/2006/06/nationalparks200606.

7   This legislation was sponsored and carried through the Senate by Republican Rob Portman.

8   The 2018–2019 federal government shutdown under President Trump lasted thirty-five days and cost the government over $5 billion.

9   See https://www.c-span.org/video/?315631-1/national-park-service-government -shutdown.

10   Best Places to Work in the Federal Government: https://bestplacestowork.org/rankings /detail/IN10#tab_category_tbl.

Chapter 7

1   Linda Bilmes and John Loomis, *Valuing U.S. National Parks and Programs: America's Best Investment* (New York: Routledge Press, 2019).

2   See https://www.si.edu/regents.

3   Visitor Experience Improvements Authority, chap. 1019 of title 54, United States Code, Public Law 114-289, section 701.

BIBLIOGRAPHY AND
FURTHER READING

The following list includes books cited in the manuscript as well as our selection of essential reading on the history of the national parks.

Advancing the National Park Idea. National Parks Second Century Commission Report. https://www.npca.org/resources/1900-national-parks-second-century-commission-report.

Albright, Horace. *Creating the National Park Service: The Missing Years.* Norman: University of Oklahoma Press, 1999.

Bilmes, Linda, and John Loomis. *Valuing U.S. National Parks and Programs: America's Best Investment.* New York: Routledge Press, 2019.

Brennan, Summer. *The Oyster War: The True Story of a Small Farm, Big Politics, and the Future of Wilderness in America.* Berkeley, CA: Counterpoint, 2015.

Butler, Mary Ellen. *Prophet of the Parks: The Story of William P. Mott, Jr.* Ashburn, VA: National Recreation and Park Association, 1999.

Caffrey, Dr. Maria. Written Testimony before the House Committee on Natural Resources. July 25, 2019. https://naturalresources.house.gov/imo/media/doc/Dr.%20Caffrey%20-%20Written%20Testimony%20-%20FC%20Ov%20Hrg%2007.25.19%20(Scientific%20Integrity).pdf.

Cahn, Robert. *The Birth of the National Park Service: The Founding Years, 1913–1933.* Salt Lake City: Howe Brothers, 1985.

Department of the Interior. Memorandum, Options Regarding Applications for Hardrock Mineral Prospecting Permits on Acquired Lands Near a Unit of the National Park System. April 16, 1998. https://doi.opengov.ibmcloud.com/sites/doi.opengov.ibmcloud.com/files/uploads/M-36993.pdf.

Freemuth, John. *Islands under Siege: National Parks and the Politics of External Threats.* Lawrence: University Press of Kansas, 1991.

Grinnell Resurvey Project. Museum of Vertebrate Zoology, University of California, Berkeley. https://mvz.berkeley.edu/Grinnell/.

Hartzog, George. *Battling for the National Parks*. New York: Moyer Bell Limited, 1993.

Heacox, Kim. *An American Idea: The Making of the National Parks*. Washington, DC: National Geographic Society, 2001.

House Natural Resources Committee and Oversight and Government Reform Committee Joint Hearing. https://www.c-span.org/video/?315631-1/national-park-service -government-shutdown.

House of Representatives Report 94-1318, *The Degradation of Our National Parks*. June 30, 1976.

House of Representatives Report 95-106, *Protecting Redwood National Park*, Report of the Committee on Government Operations. March 23, 1977.

Ise, John. *Our National Park Policy: A Critical History*. Baltimore: Johns Hopkins Press, 1961.

Kaufmann, John. *Alaska's Brooks Range: The Ultimate Mountains*. Seattle: The Mountaineers, 1988.

Mengak, Kathy. *Reshaping Our National Parks and Their Guardians: The Legacy of George B. Hartzog*. Albuquerque: University of New Mexico Press, 2012.

National Park Foundation. *LGBTQ America: A Theme Study of Lesbian, Gay, Bisexual, Transgender, and Queer History*. 2016. https://www.nps.gov/subjects/tellingallamericansstories /index.htm.

National Park Service. *American Pacific Islander National Historic Landmark Theme Study*. 2017. https://www.nps.gov/subjects/tellingallamericansstories/index.htm.

National Park Service. Director's Order 100. Enacted December 20, 2016. Rescinded August 16, 2017.

National Park Service Centennial Act. Public Law 114-289, Title IV, National Park Foundation Authorities, Section 401.

National Park Service System Plan. https://parkplanning.nps.gov/files/NationalParkService SystemPlan2017.pdf.

National Park System Advisory Board. *American Latinos and the Making of the United States: A Theme Study*. 2013. https://www.nps.gov/subjects/tellingallamericansstories/index .htm.

National Park System Advisory Board. Report: Rethinking the National Parks for the 21st Century. Washington, DC: National Geographic Society, 2001.

National Park System Advisory Board. Revisiting Leopold: Resource Stewardship in the National Parks, 2012. https://www.nps.gov/calltoaction/PDF/LeopoldReport_2012.pdf.

National Parks and Recreation Act of 1978. P.L. 95-625. https://www.govinfo.gov/content /pkg/STATUTE-92/pdf/STATUTE-92-Pg3467.pdf#page=72.

Obama, Barack. Remarks by the President at the Dedication of the César E. Chávez National Monument. National Archives, 2012. https://obamawhitehouse.archives.gov/the-press -office/2012/10/08/remarks-president-dedication-cesar-chavez-national-monument -keene-ca.

Public Land Law Review Commission. Government Printing Office. June 20, 1970.

The Recreation Imperative: A Draft of the Nationwide Outdoor Recreation Plan Prepared by the Department of the Interior. Henry M. Jackson, chair, Committee on Interior and Insular Affairs, US Senate. September 1974.

Rettie, Dwight. *Our National Park System: Caring for America's Greatest Natural and Historic Treasures*. Urbana: University of Illinois Press, 1999.

Roberts, Malcomb. *The Wit and Wisdom of Wally Hickel.* Anchorage, AK: Searchers Press, 1994.

Rothman, Hal. *Preserving Different Paths: The American National Monuments.* Urbana: University of Illinois Press, 1989.

Sellars, Richard. *Preserving Nature in the National Parks.* New Haven, CT: Yale University Press, 1997.

Shankland, Robert. *Steve Mather of the National Parks.* New York: Alfred A. Knopf, 1951.

Shnayerson, Michael. Who's Ruining Our National Parks? *Vanity Fair.* https://www .vanityfair.com/news/2006/06/nationalparks200606.

Swain, Donald C. *Wilderness Defender: Horace M. Albright and Conservation.* Chicago: University of Chicago Press, 1970.

36 CFR Parks, Forests and Public Property, Part 2. Resource Protection, Public Use and Recreation, §2.1 Preservation of Natural, Cultural and Archeological Resources. Government Printing Office.

Tweed, William. *Uncertain Path: A Search for the Future of National Parks.* Berkeley: University of California Press, 2010.

Visitor Experience Improvements Authority. Public Law 114-289, Title IV, Sec. 701. Chapter 1019 of Title 54, United States Code.

Wirth, Conrad. *Parks, Politics and the People.* Norman: University of Oklahoma Press, 1980.

Wauer, Roland. *My Wild Life.* Lubbock: Texas Tech University Press, 2014.

Wright, George Melendez. *Fauna Series No. 1.* National Park Service, 1932.

Yard, Robert Sterling. *The Book of the National Parks.* New York: Charles Scribner's Sons, 1919.

INDEX